PEER-LED TEAM LEARNING

ORGANIC CHEMISTRY

THE WORKSHOP PROJECT

SPONSORED BY THE NATIONAL SCIENCE FOUNDATION

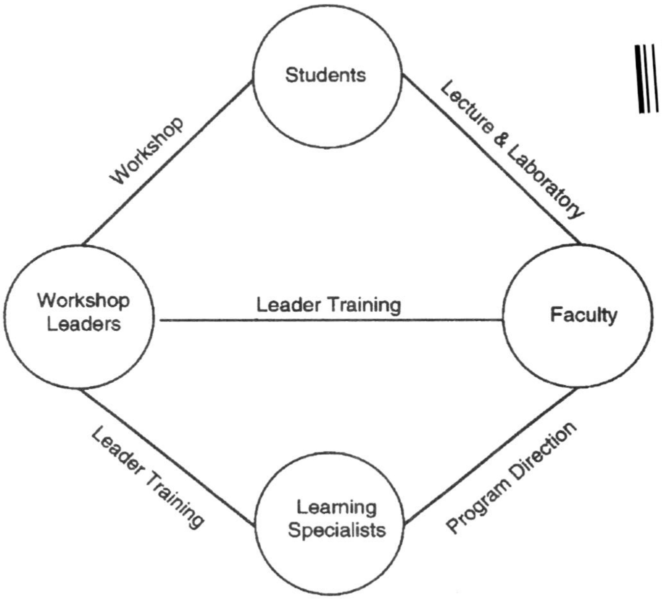

J. A. KAMPMEIER
PRATIBHA VARMA-NELSON
DONALD K. WEDEGAERTNER

PRENTICE HALL SERIES IN EDUCATIONAL INNOVATION

Prentice Hall

PRENTICE HALL, Upper Saddle River, NJ 07458

Executive Editor: ***John Challice***
Project Manager: ***Kristen Kaiser***
Editorial Assistant: ***Gillian Buonanno***
Special Projects Manager: ***Barbara A. Murray***
Production Editor: ***Benjamin St. Jacques***
Manufacturing Manager: ***Trudy Pisciotti***
Manufacturing Buyer: ***Michael Bell***
Supplement Cover Manager: ***Jayne Conte***
Supplement Cover Designer: ***Maureen Eide***

Printed in the United States of America

10 9 8 7 6 5 4 3 2 1

ISBN 0-13-028413-0

Prentice-Hall International (UK) Limited, *London*
Prentice-Hall of Australia Pty. Limited, *Sydney*
Prentice-Hall Canada, Inc., *Toronto*
Prentice-Hall Hispanoamericana, S.A., *Mexico*
Prentice-Hall of India Private Limited, *New Delhi*
Prentice-Hall (Singapore) Pte. Ltd.
Prentice-Hall of Japan, Inc., *Tokyo*
Editora Prentice-Hall do Brasil, Ltda., *Rio de Janeiro*

Preface to the Peer-Led Team Learning Series

The Workshop Chemistry Project was an exploration, development, and application of the concept of peer-led team learning in problem-solving Workshops in introductory chemistry courses. A pilot project was first supported by the National Science Foundation, Division of Undergraduate Education, in 1991. In 1995, the Workshop Chemistry Project was selected by NSF/DUE as one of five systemic initiatives to "change the way introductory chemistry is taught." In the period 1991-1998, the project grew from the initial explorations at the City College of New York to a national activity involving more than 50 faculty members at a diverse group of more than 30 colleges and universities. In 1998-1999, approximately 2500 students were guided in Workshop courses by 300 peer leaders per term. In Fall 1999, NSF chose the Workshop Project for a National Dissemination Grant to substantially broaden the chemistry participation and to extend the model to other SMET disciplines, including biology, physics and mathematics.

Peer-Led Team Learning: A Guidebook is the first of a series of five publications that report the work of the Project during the systemic initiative award (1995-1999). The purpose of these five books is to lower the energy barrier to new implementations of the model. The *Guidebook* is a comprehensive account that works back and forth from the conceptual and theoretical foundations of the model to reports of "best-practice" implementation and application. Three other books provide specific materials for use in Workshops: *General Chemistry*; *Organic Chemistry*; and *General, Organic and Biochemistry*. One book in the series, *On Becoming a Peer Leader*, provides materials for leader training.

The collaboration of students, faculty, and learning specialists is a central feature of the Workshop model. The project has been enriched by the talents and energy of many participants. Some of their names are found throughout these books; many others are not identified. In either case, we are most grateful to all those who have advanced the model by their keen insight and enthusiastic commitment.

We also acknowledge, with pleasure, the support of the National Science Foundation, NSF/DUE 9450627 and NSF/DUE 9455920. Our work on the second NSF award was skillfully guided by our National Visiting Committee, Michael Gaines, Chair; Joseph Casanova; Patricia Cuniff; David Evans; Eli Fromm; John Johnson; Bonnie Kaiser; Clark Landis; Kathleen Parson; Arlene Russell; Frank Sutman; Jeffrey Steinfeld; and Ronald Thornton; we value their advice and encouragement. The text of the *Guidebook* was repeatedly processed by Arlene Bristol, with exceptional skill and remarkable patience. Finally, we appreciate the vision and commitment of John Challice and Prentice Hall to make this work readily available to a large audience.

Books are written for you, the readers. We welcome your comments and insights. Please contact us at the indicated e-mail addresses.

The Editors, Fall 1999

David K. Gosser	gosser@scisun.sci.ccny.cuny.edu
Mark S. Cracolice	markc@selway.umt.edu
J. A. Kampmeier	kamp@chem.rochester.edu
Vicki Roth	vrth@mail.rochester.edu
Victor S. Strozak	vstrozak@gc.cuny.edu
Pratibha Varma-Nelson	varmanelson@sxu.edu

Introduction to Workshops for Organic Chemistry

Workshop Chemistry helps students acquire the skills, experience, and habits of study and thought required to be successful in organic chemistry. Accordingly, the purpose of this collection of organic Workshop problems is to lower the activation energy for faculty to implement Workshops. We have collected here Workshop problems that have worked well in our courses. We invite users of this collection to appropriate freely and modify as necessary to assemble Workshop materials for their own courses. The problems are available in hard copy and on disk (Microsoft Word 98 and ChemDraw for the Mac) to make it easy and fast to assemble sets of problems.

There is some truth to the idea that "problems are neither good nor bad; it depends on how you work with them." As a result, some of the Workshop problems will look familiar. Structure proofs are prime examples of traditional problems that work exceedingly well in the Workshop format. We have modified the traditional problem to make explicit the Observation-Deduction reasoning that is required to solve the problem. This simple device provides a mechanism for a stepwise participatory problem-solving by the group. It is also a prime example of the power of "making thinking visible."

However, it is not true that all problems can be made into good Workshop problems. Our intention is to provide a collection of Workshop problems and not just another organic problem book. In general, we have not included drill problems because they tend to be very direct: question and answer. The Workshop emphasis is not on answers; rather, it is on the habits of analysis, thought, evaluation, and discussion that lead to the formulation and recognition of answers. For this reason we do not provide answer keys to either students or Workshop leaders. We want students to learn to rely on their own judgments. Good Workshop problems provoke questions and debate and provide many opportunities to check conceptual understanding. Challenging problems do that; drill problems do not.

The best Workshop problems draw the students into **active** engagement with the material and with each other. The challenge helps engage students in both dimensions by setting a goal that is not immediately attainable by individual effort. To be effective, the magnitude of the challenge must be just right. If the stretch is too small, the challenge is too small and the students will not need to learn. If the stretch is too great, the challenge is frustrating, and the students will give up without learning. The right challenge is attainable with some time, thought, and discussion by the group. Some problems may look like drill but actually have quite different objectives. Beginning students often have difficulty recognizing recurring ideas. Problem 4 in Workshop 5 is designed to help students get beyond isolated equations to a unified, connected view of a group of reactions. Starburst problems (Workshop 8) emphasize the need to master the empirical data base in organic chemistry, a requirement that comes as a shock to many students. Many Workshop problems look like hard exam problems for two reasons. First, the resemblance is motivating and meets the test of relevance. Second, we cribbed from our old exams.

Students need to prepare to do challenging problems. Reading the text, studying the lecture, and doing the drill problems in the textbook are prerequisite to full participation in a Workshop. Students who follow that routine will do best. Most students will look over the Workshop ahead of time to guide their preparation. We expect students to do their homework, but we do not expect them to do the Workshop problems ahead of time. Workshop problems are not homework.

A Workshop works best if only one or two key ideas are explored in coherent, connected fashion. A good Workshop, therefore, is not just a willy-nilly collection of isolated problems. A drill problem might even work well to help students get started. Ultimately, the component problems in a Workshop should build on one another to help develop a deeper understanding of a **big idea**. In this book, we provide a full set of Workshops for a typical one-year organic course.

The book is designed for teachers. We expect that these specific examples of complete Workshops, accompanied by commentaries, will be useful to faculty who are just getting started with Workshops. Workshops are analyzed in some detail to show what we mean by **big ideas** and **take-home points** and to illustrate how an individual Workshop builds from an accessible starting point to a sophisticated outcome. Different kinds of Workshops illustrate different techniques. The Workshops focus on "trouble spots" and are built around ideas to help students get through the difficulties. The gloss often includes suggestions about how to lead the Workshop. The ideas discussed are the kinds of issues that we review with our leaders before they lead the Workshop. Finally, we cannot overemphasize that we do not discuss *answers* with the leaders. The fundamental concept is that Workshops are about constructing and recognizing answers. Even the most carefully constructed Workshop can be destroyed by students, teachers or leaders who surrender to "the Answer to #3 is ...."

We faced some hard decisions in assembling these Workshop problems. The most serious challenge was that we teach our subject in different sequences, using different books and with different objectives. In addition, we also emphasize the intellectual coherence of the subject by repeatedly using ideas that we have developed previously. As a result, it may be hard to transfer an intact Workshop from one course to another. In the end, we decided to put the Workshops together "our way," recognizing that users will undoubtedly add, subtract, rearrange, and rewrite according to their specific needs.

We also puzzled about the proper way to treat spectroscopy. Some faculty make it part of the lecture course, some cover the topic in laboratory, and some do both. In the end, we decided that there are many excellent tutorials and collections of spectroscopic problems and that we would simply use spectroscopy as a working tool in our Observation-Deduction problems.

This collection is a group effort, involving different faculty working with different texts, in different schools, and with different students. The contributing faculty were Madeline Adamczewski, San Jose City College; Joseph Dinnocenzo, University of Rochester; Joshua Goodman, University of Rochester; Jack Kampmeier, University of Rochester; Andrew S. Kende, University of Rochester; Bryan Roberts, University of Pennsylvania; Pratibha Varma-Nelson, St. Xavier University (Chicago), and Donald Wedegaertner, University of the Pacific.

We welcome your corrections and advice. The best method is to write directly on the text and return the annotated copy to J. A. Kampmeier at the end of the semester. We will send you a fresh copy. You may also e-mail or fax more general comments. We are grateful to our students for good advice and are particularly interested in comments from your students and Workshop leaders.

Ms. Arlene Bristol provided marvelous skill and forbearance while assembling these materials from disparate sources. Please contact her directly if there are technical questions about the format of the disk.

Mailing Addresses:
Professor J. A. Kampmeier *or*
Ms. Arlene Bristol
Department of Chemistry
University of Rochester
River Campus P. O. Box 270216
Rochester NY 14627-0216

E–mail Addresses:
kamp@chem.rochester.edu.
bristol@chem.rochester.edu

Fax Number:
(716)-473-6889

Table of Contents

WORKSHOP

Week 1

Structure: Functional Groups

1. An experimental technique called ^{13}C nuclear magnetic resonance spectroscopy (NMR) allows chemists to tell how many different kinds of carbons there are in a molecule and whether carbons are primary (1°), secondary (2°), tertiary (3°), or quaternary (4°). Give Kekulé structures (i.e., use lines for electron pair bonds) for the following compounds having molecular formula C_6H_{12}. Also, on each structure, identify carbons as 1°, 2°, 3°, or 4°; tell how many different kinds of carbons there are; and designate which carbons are equivalent.

   a. A compound having only single bonds and only secondary carbons

   b. A compound having only single bonds and primary, secondary, and tertiary carbons

   c. A compound having only single bonds and only primary, secondary, and quaternary carbons

   d. A compound having only single bonds and primary, secondary, tertiary, and quaternary carbons

2. The following isomers of molecular formula C_5H_{12} were treated under appropriate conditions to give all possible monochloro derivatives, $C_5H_{11}Cl$, in which H's are replaced by Cl. These $C_5H_{11}Cl$ constitutional isomers, derived from each C_5H_{12} compound, could be separated either by careful fractional distillation or by gas chromatography (GC). Give the structures of the following C_5H_{12} compounds and the $C_5H_{11}Cl$ derivatives indicated. Also, name each of these compounds.

   a. A C_5H_{12} compound that gives three $C_5H_{11}Cl$ derivatives

   b. A C_5H_{12} compound that gives four $C_5H_{11}Cl$ derivatives

   c. A C_5H_{12} compound that gives only one $C_5H_{11}Cl$ derivative

   d. Also give the structure of a C_8H_{18} compound that gives only one $C_8H_{17}Cl$ derivative. Give the structure of the derivative as well.

3. Draw Kekulé structures (show all bonds as lines and show all nonbonding electron pairs) for constitutional isomers with molecular formula $C_3H_6O_2$. Make sure that the following functional groups are included in these isomers: carboxylic acid, ester, ether, aldehyde, ketone, alcohol. Circle each functional group and indicate its appropriate name.

4. A drug company touted its discovery of a new drug, having a molecular formula of $C_{15}H_{33}$, that was supposed to be effective for treating just about any known disease. Ignoring the extravagant curative claims, discuss whether the formula for this drug is reasonable and whether you would invest in this company.

WORKSHOP GUIDE

Week 1

Structure: Functional Groups

It will help beginners and those less confident if they are eased into a problem. Consider (in advance) how to break the problem into smaller steps and then use that framework to guide the students' discussion with one another. For example:

Problem 1: The first step in any problem is to figure out what is being asked; that is, What is the question? A part of this step is understanding the text. What does 1°, 2°, 3°, and 4° mean?

The second step is to figure out where to start.

A leader could break out these two tasks in order to get started. The students can talk among themselves for a while, focusing on these two key steps. When the students are ready, the leader might ask for an open discussion of the first two steps. This should bring out the definitions of different kinds of carbons, the meaning of equivalency, and the fundamental problem that they are to find structures that are consistent with specific observations. Ideally, the students will see that the place to start is with the molecular formula. Why? The formula applies equally to all the specific problems. More fundamentally, specific structures have specific formulas; that is, the formula excludes (limits) possible type of structures (**big idea**). After the students work through to the conclusion that C_6H_{12} must be a ring or a double-bond structure, the specification that the structures have only single bonds leads to a **big conclusion**: the structures have a ring. The leader must not tell the students this. The leader's job is to lead them to that discovery.

Now the group is probably ready to start to try to find specific structures that meet the specified criteria (observations) by working and talking among themselves. After a while, when the students are ready, the leader can facilitate an open discussion of their solutions (structures). The leader should try to get students to defend particular structures and to challenge other structures. The students should work to consensus about the validity of any particular structure. *The leader may not stand up and give answers or respond to particular structures with yes-no or right-wrong.* The leader may (and should) comment on various **take-home points** that will not be obvious to the students such as the following:

- Structures may be the same even though they look different (are drawn differently). This idea is elucidated better in *Problem 2.*

- Observations are best used to exclude structures. This is rigorous. There may be more than one structure that is consistent with the observations; parts 1b and 1c have several solutions. The leader can start a discussion about the difference between necessary and sufficient reasons.

- There is a logic to the problems: as the carbons get more substituted (or branched) the rings must get smaller. One set of solutions proceeds very systematically: (a) zero branches (six-carbon ring), (b) one branch (five-carbon ring), (c) two branches (four-carbon ring) and (d) three branches (three-carbon ring).

In the absence of guidance, students will read rapidly through the problem and start scribbling structures, in more or less random fashion. By imposing a stepwise process, the leader can guide them to the logic of solving the problem.

Problem 2: The leader can proceed as in *Problem 1*. What is the question? Where shall we start? *Derivative* means that the particular $C_5H_{11}Cl$ is derived (is formed) from a particular C_5H_{12}.

Part *2(d)* is best held for "extra credit" at the end of the Workshop. The initial focus should be on *2(a)*, *(b)*, and *(c)* which are closely connected.

The starting point is the formula C_5H_{12}. What does it mean? The leader might get someone to write out any C_5H_{12}, then get someone to make that C_5H_{12} into a $C_5H_{11}Cl$, then consider whether that C_5H_{12} could give more than one $C_5H_{11}Cl$. The students will then be ready to find specific solutions to specific problems. The leaders must let them work away.

Take-Home Points. This is a great problem to emphasize that structures may look different but be the same. The leader should bring a model set and have students build two seemingly different, but identical, molecules and compare models. Because a carbon atom with four bonds is tetrahedral and because rotation around σ bonds is easy, more groups are equivalent than might seem at first glance.

Problem 1 was about different carbons; *Problem 2* is about different hydrogens in a molecule. Why not write C_5H_{12} with zero branches, then C_5H_{12} with one branch, then C_5H_{12} with two branches. Are there any more? With a given C_5H_{12}, replace the different H's one at a time, in systematic fashion to generate the possible $C_5H_{11}Cl$ structures.

Problem 3: As before, What is the problem? What are we supposed to do? How do we start?

Take-Home Points. The rules of valence and the Lewis structural (octet) theory give us a way to construct different structures for different molecules (isomers), even though the molecular formula is the same. The functional group idea focuses attention on the structural unit that is changing. Different molecules (different functional groups) have specific names. They also have different properties (of which the name is one) because they have different structures.

Problem 4: For extra-credit fun, just like 2(d). If the students did *Problems 1* and *2*, they can now do *2(d)* and *4*, which might look hard at first.

Wrap Up. It is very important to pace the Workshop so that the students give equal time to *Problems 1, 2* and *3*. It is not necessary to finish every part of *1*, *2*, and *3*, once the ideas are clear. It is much better to skip *1(d)* or *2(c)* and move on to *Problem 3* than to not get to *Problem 3* at all.

Finally, the leader might plan a short (10 minutes) summary of the **take-home points**. Get the students to identify the **take-home points**. The leader can work as scribe. This is a good way to wrap up.

WORKSHOP

Week 2

Structure: Bonding

1. a. Draw Kekulé structures for each of the following compounds. Be sure to show all bonding and nonbonding valence electrons. Also, clearly indicate any formal charges on atoms.

   b. Consider the molecular geometry information given for each of the following compound using this information, specify the orbitals that each atom could use in σ– and π–bonding (*sp*^{2}, *sp*^{3}, *p*, etc.) and for holding nonbonding electron pairs. Explain how your orbital assignments are consistent with the observed geometries.

   DCH=CHBr (1–bromo–2–deuterioethylene)
   (two isomers)
   (all atoms are coplanar)

   $CH_3CO_2^-$ (acetate ion)
   HCC ∡ 109°
   ∡ CCO = OCO = 120 °

   H_2CO (formaldehyde)
   HCO ∡ = 120°

   $(CH_3)_2SO$ (DMSO)
   ∡ HCS = CSO = CSC = 109°

2. Three dicarboxylic acids, $C_2H_2(CO_2H)_2$, **I**, **J**, and **K**, are catalytically hydrogenated (reacted with H_2 in the presence of a catalyst) to give dicarboxylic acids with formulas of $C_2H_4(CO_2H)_2$. Hydrogenation of both **I** and **J** gives the same dicarboxylic acid **L**. Compound **K** hydrogenates to form compound **M**. Give structures for compounds **I-M**. **Explain your reasoning.**

   Observation **Deduction**

3. Two isomeric compounds, **A** and **B**, are known to each have a monosubstituted benzene ring (C_6H_5–). Both have the formula $C_6H_5C_3H_5O_2$ and both are insoluble in water. However, when they are treated with dilute aqueous NaOH, **A** dissolves, but **B** does not. Give structures for **A** and **B** *consistent* with this information. **Explain your reasoning.**

   Observation **Deduction**

4. Although we discussed the idea of tetrahedral carbon as proposed by van't Hoff and Le Bel, square planar carbon is also a possible structure. Show how this structure can be eliminated by considering CH_2Cl_2 and CH_2BrCl as examples of disubstituted methanes.

a. How many isomers would be possible in each instance if the carbon had a square planar structure?

b. How many isomers are possible in each instance if the carbon is tetrahedral?

Consider CHBrClF as an example of a trisubstituted methane.

c. How many isomers would be possible if the carbon atom were square planar?

d. How many isomers are possible for CHBrClF if carbon is tetrahedral?

The experimental observations are:

CH_2Cl_2	one isomer
CH_2BrCl	one isomer
CHBrClF	two isomers

WORKSHOP GUIDE

Week 2

Structure: Bonding

Problem 1: Round robin is a great technique for getting everyone involved. It subdivides the work so that individual tasks (steps) are manageable. By aggregating the individual roles, the group solves the problem.

The Workshop leader randomly assigns students to the tasks (or steps). The group then follows the steps to the solution. The leader can act as scribe. One member of the group can summarize the structure. For the next structure, the assigned tasks are passed to the person on the right. The process continues until every one has had a chance at several tasks. Here is a round robin approach to molecular structure.

Step 1: Start with the formula. Decide how the atoms are connected. If there are several isomeric structures, choose one with which to proceed.

Step 2: Write the Lewis structure.

Step 3: Consider the geometry, which is either (a) given or (b) guessed from VSEPR, for example, $CH_3C{\equiv}N$.

For the "central atom" of the methyl carbon, the number of electron pair units is 4. For the "central atom" of the nitrile group, the number of electron pair units is 2.

Step 4: Overlap hybrid orbitals of the central atom with appropriate orbitals of the peripheral atoms, making the appropriate number of σ bonds.

Step 5: Make π bonds as required to accommodate multiple bonds.

Step 6: Distribute the nonbonding electron pairs to the appropriate atoms.

Step 7: Check that octets are filled and that electrons are paired. Calculate formal charges.

Step 8: Label bonds as σ or π, identifying the orbitals that overlap to make the bonds: for example, $\sigma(\mathrm{H}1s\mathrm{C}sp^3)$.

Take-Home Points. There is a logical method for writing structures. It is essential to identify the non-bonding pairs; they are centers of reactivity because they are available for bonding. π-Bonding always involves *p*-orbitals on C, O, or N. The choice of hybrid orbitals to use follows from the geometry.

Problems 2 and *3:* Puzzle problems are great Workshop problems. Here is one way to proceed. Ask the students to read the problem carefully to find out what they are supposed to do (figure out structures that are consistent with the given observations and the rules of structure). The leader can volunteer to serve as scribe. Ask the students to reread the problem to find the observations. When the students are ready, ask them for the observations, one by one. Write them on the board. (This is important because everyone can participate.) Now, take the observations, one by one, and ask the students to tell what can be deduced about structure from each observation. (Pick respondents carefully, keeping the better students for the tougher observations.) Then, encourage the group to collect the deductions to make superdeductions, and so on, until the problem is solved.

Beginning students tend to try to solve structure problems all at once, taking all of the data in one big gulp. They also tend to jump to conclusions. They guess a structure on the basis of one observation, for example, ($C_5H_{12} = CH_3(CH_2)_3CH_3$), and then will not let go of it even when the structure is excluded by subsequent observations. It is always much better to have multiple hypotheses (there are three C_5H_{12}'s); then it is psychologically easy to exclude hypotheses.

Take-Home Points. Solving problems such as these requires students to know the empirical facts (this structure has these properties or vice versa) and to connect the facts in a purposeful manner. Connecting individual deductions to make larger deductions is both logical and creative. It is distinctly higher order thinking than knowing the simple observation-deduction pairs.

The problem may not have a unique solution. The solution must be consistent with the observations, but the observations may not be sufficient. It is important to recognize ambiguity, that is, what you know for sure and what you do not know. On the other hand, when the data are necessary and sufficient, then there is a unique answer and you can "bet your life on it." This always comes as a surprise to students. They want to say "it is probable," and "it might be"; they are reluctant to say "it is." For example, there is only one C_6H_{12} that has only single bonds and only secondary carbons (Workshop 1).

Problem 4: This would be a good problem to solve by dividing the group into teams. Students can count isomers and then compare and debate answers.

This is a great Workshop problem because it gets directly at epistemological issues. How is it that we know what we think we know? This problem emphasizes that there is a logical, rigorous way of eliminating the hypothesis that tetravalent carbon has a square planar geometry. The tetrahedral hypothesis is shown to be consistent with observation. The logic does not prove that tetravalent carbon is tetrahedral, but it does prove that it is not square planar.

Take-Home Point. We think what we think because of a trail of logic, not because somebody said it was so.

WORKSHOP

Week 3

Structure: Properties: Acids and Bases

1. For each of the following reactions, designate the acids and the bases and use curved arrows to show the flow of electrons as the reaction proceeds from left to right. Circle the major species at equilibrium. Start by drawing complete Kekulé structures showing all bonding and nonbonding electron pairs at the reaction centers.

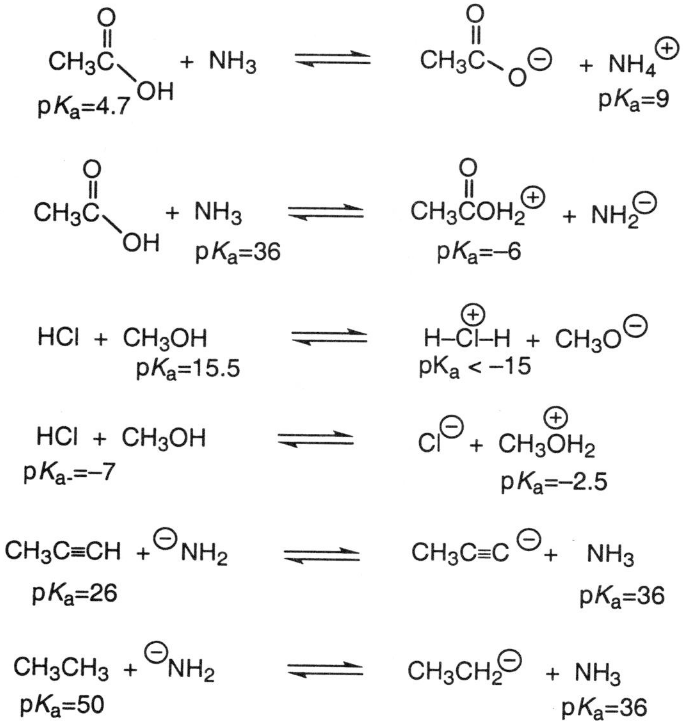

2. Draw complete Kekulé structures of reactants and products using lines for bonds, showing nonbonding electron pairs, and indicating formal charges on atoms. Identify the acid and the base for each reaction. Use curved arrows to show the movement of electron pairs as the reactions proceed from left to right.

$$CH_3SCH_3 + BF_3 \rightleftharpoons (CH_3)_2\overset{\oplus}{S}\text{–}\overset{\ominus}{B}F_3$$

$$^{\oplus}CH_3 + H_2O \rightleftharpoons CH_3\overset{\oplus}{O}H_2$$

$$CH_3O^{\ominus} + CH_3SH \rightleftharpoons CH_3OH + CH_3S^{\ominus}$$

$$CH_3\overset{\overset{O}{\|}}{C}CH_3 + H_3O^{\oplus} \rightleftharpoons H_2O + CH_3\overset{\overset{\oplus OH}{\|}}{C}CH_3$$

$$HC{\equiv}C^{\ominus} + H_2C{=}O \rightleftharpoons HC{\equiv}C\text{–}CH_2\text{–}O^{\ominus}$$

$$CH_2{=}CH_2 + HCl \rightleftharpoons {}^{\oplus}CH_2\text{–}CH_3 + Cl^{\ominus}$$

$$I^{\ominus} + CH_3Br \rightleftharpoons ICH_3 + Br^{\ominus}$$

3. Predict the product(s) for the following reactions. The ideas that you used in the previous problems will be useful.

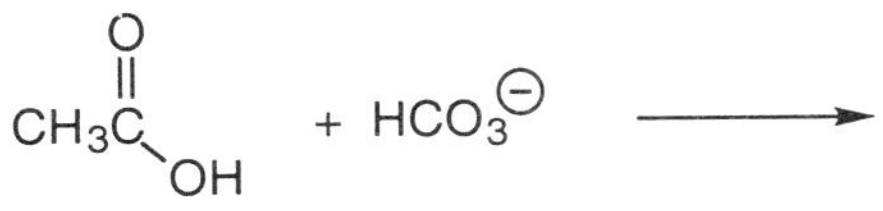

$$B(CH_3)_3 + (CH_3)_3N \longrightarrow$$

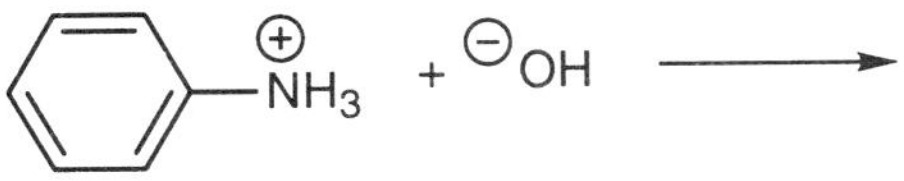

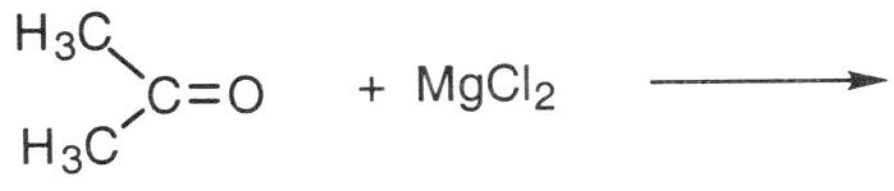

$$H_2SO_4 + CH_3OCH_3 \longrightarrow$$

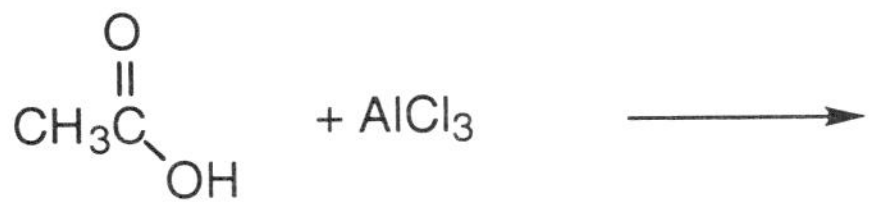

4. For each of the following pairs, one reaction proceeds to the specified products and the other does not. Predict which is which and *clearly explain your choice.* It will help to start with curved arrows to show which bonds are made and which bonds are broken in each reaction.

a. $HO^{\ominus} + Br\text{–}CH_3 \longrightarrow HO\text{–}Br + {}^{\ominus}CH_3$

$HO^{\ominus} + Br\text{–}CH_3 \longrightarrow HO\text{–}CH_3 + Br^{\ominus}$

b. $H\text{–}C{\equiv}C^{\ominus} + H_2C{=}O \longrightarrow H\text{–}C{\equiv}C\text{–}CH_2\text{-}O^{\ominus}$

$H\text{–}C{\equiv}C^{\ominus} + H_2C{=}O \longrightarrow H\text{–}C{\equiv}C\text{–}O\text{–}CH_2^{\ominus}$

c, ${}^{\ominus}NH_2 + H\text{–}OCH_3 \longrightarrow H_2N\text{–}OCH_3 + {}^{\ominus}H$

${}^{\ominus}NH_2 + H\text{–}OCH_3 \longrightarrow H_2N\text{–}H + {}^{\ominus}OCH_3$

WORKSHOP GUIDE

Week 3

Structure: Properties: Acids and Bases

This Workshop is an example of a carefully structured development from the familiar (proton transfer) to the unfamiliar (predicting products of new reactions). There is also a structured progression from equilibrium concepts in *Problem 1* to rate concepts in *Problem 4*. The switch from equilibrium (first-year chemistry) to rate (second-year chemistry) is a stumbling block for many students. Drawing complete Kekulé structures, as directed in *Problem 1*, is the proper starting point for all four of these problems. Using curved arrows to show the flow of electron pairs is an important aid to understanding reactions and is an important prelude to their use in complex mechanistic problems.

Problem 1: If students have a problem getting started, the leader might ask them to identify which species are giving up (donating) protons (acid) and which are taking on (accepting) protons (base).

Take-Home Point. Proton transfer occurs from one electron pair to another. Think of the equilibrium as a competition for the proton; the weaker acid wins the competition (predominates at equilibrium).

Problems 2 and *3:* These problems are collections of Bronsted and Lewis acid-base reactions. The **big idea** is to help students see that proton transfers belong to a subset of Lewis donor-acceptor interactions. Work with them to find the electron pair donor and acceptor character of all the reactions. That is, of course, the basis for predictions in *Problem 3. Problem 3* is a very big intellectual and psychological jump for students because they must make predictions (go out on limbs).

Problem 4: For *Problem 4*, the **big idea** is that the more favorable reaction involves transferring negative charge to the more electronegative atom. *Problem 4* introduces the idea of competitive reactions without saying much about it. The problem looks ahead to the transition from thinking about equilibrium to thinking about rate. We will return to this issue in full form in Workshop 5.

Take-Home Points. This Workshop addresses a major dilemma for students: How does one know what reacts with what? How can one predict products?

WORKSHOP

Week 4

Conformational Stereochemistry: Alkanes and Cycloalkanes

1. Consider rotations around the C_2–C_3 bond of 2–methylbutane. Build molecular models of all the staggered conformations. Draw the conformations using Newman projections. Circle the low energy conformation(s). **Explain your choices.** Construct a qualitative plot of energy versus angle of rotation around the C_2–C_3 bond.

2. a. For butane, make molecular models of the anti and gauche conformations as well as the conformation in which the two methyls are eclipsed. Using your molecular model, measure the distance between the methyl carbons in the anti, gauche, and eclipsed conformational isomers. Make careful drawings (sawhorse) of the three conformational isomers.

   b. Make molecular models for methylcyclohexane and carefully draw the two chair conformations. Measure the distance between the methyl carbon and the carbons at the 3– and 5– positions of the ring. Compare these measurements with those that you previously made on butane. On the basis of your measurements, predict which of the two cyclohexane conformations will predominate at equilibrium. Is the methyl group axial or equatorial in the more stable conformation? **Explain your reasoning.**

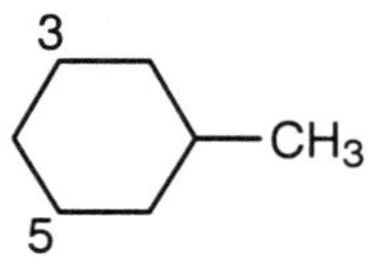

3. Draw all isomeric dimethylcyclobutanes, clearly showing stereochemistry. Designate as cis or trans where appropriate. Indicate whether structures are related as constitutional isomers or stereoisomers.

4. Using only chair conformations, make and draw molecular models for all the 1,4–dibromocyclohexanes. Clearly indicate which are related as conformational isomers and which are related as configurational isomers. Designate the configurational isomers as cis or trans. **Explain** which isomers equilibrate rapidly with each other (interconvert).

WORKSHOP GUIDE

Week 4

Conformational Stereochemistry: Alkanes and Cycloalkanes

The leader might assign students to work on these problems in pairs or in small groups. The Workshop leader can go from group to group to monitor progress and to offer suggestions. After all have worked through a particular problem, it is good to bring the whole group together to compare, summarize, and discuss results. Be sure that students build molecular models before working with 2-D drawings on paper. Having built models in *Problems 1* and *2*, it should be clear to them that models could also help with *Problems 3* and *4*. Be prepared to make this suggestion if necessary.

Workshop 4 is quite different in character and learning style from its predecessor. Whereas Workshop 3 made use of an abstract theory to analyze and predict, Workshop 4 is concrete and physical. Students can hold the models in their hands, see with their eyes and measure distances. In the end, of course, the model is also an abstraction.

Important Take-Home Points:

a. Conformational isomers are interconverted by rotation around carbon-carbon single (sigma) bonds.

b. Intramolecular distances vary with conformation.

c. Examination of changes in van der Waals repulsions can be used to qualitatively predict relative conformational energies and, therefore, relative conformational populations at equilibrium.

d. All structural and stereoisomers can be accounted for by *systematically* varying the position of the methyl groups. (*Problem 3*)

e. Cis- and trans-diastereomers (and constitutional isomers) do not interconvert by rotations around single bonds.

f. The models for methylcyclohexane are not methylcyclohexane. Methylcyclohexane is, after all, a colorless liquid!

WORKSHOP

Week 5

Alkenes: Electrophilic Addition Mechanism: Carbocations

1. Draw the two chair conformations of *trans*–1,4–dichlorocyclohexane. Construct a qualitative plot (reaction energy diagram) that describes the energy of the system as a function of the progress of the interconversion of one conformational isomer into the other. Be sure to clearly indicate the coordinates on the diagram that correspond to the different conformations. Specify the energy difference that controls the population of the two conformations at equilibrium. Also, specify the energy differences that control the rate of the equilibration reactions (there is both a forward and a back reaction rate). Using a model of 1,4–dichlorocyclohexane, **explain (in words)** how your diagram corresponds to changes in the molecular structure.

2. Give the product for the following reaction and write a detailed stepwise description (mechanism) for the transformation. Identify each species in the mechanism as a Lewis acid or Lewis base and use curved arrows to show the movement of electron pairs. Construct a plot that describes the energy of the system as a function of the progress of the reaction (reaction energy diagram). On your diagram, clearly label the positions of the reactants, any intermediate(s), and the product. Also, specify the energy difference that corresponds to the ΔG° for the overall reaction and the energy difference that corresponds to $\Delta G^{\neq}$ for each step. **Explain (in words)** what is happening as the system makes its way from reactants to products.

$$H_2C{=}CH_2 + HCl \rightarrow$$

3. The addition of HBr to 2–methylpropene can, in principle, give either 1–bromo–2–methylpropane or 2-bromo-2-methylpropane. In practice, only the latter is formed. Show mechanisms for the formation of each of these alkyl bromides, using curved arrows to indicate the movement of electron pairs. Also, construct reaction energy diagrams that describe the energy of the system as a function of the progress of the formation of each alkyl bromide. Use these reaction coordinate diagrams to **explain** why 2-bromo-2-methylpropane is the only product.

4. a. Consider the first two reactions in the following set of reactions. Explain how a two-step mechanism by way of a reactive intermediate explains the formation of different products under different reaction conditions.

b. Make a table listing the electrophile and the nucleophile in each step in each of the reactions in the set.

c. Write a single, generalized mechanism that explains the entire set of reactions.

$RCH{=}CH_2 + Cl_2 \rightarrow RCHClCH_2Cl$

$RCH{=}CH_2 + Cl_2$ (in H_2O) $\rightarrow RCH(OH)CH_2Cl$

$RCH{=}CH_2 + Br_2 \rightarrow RCHBrCH_2Br$

$RCH{=}CH_2 + Br_2$ (in CH_3OH) $\rightarrow RCH(OCH_3)CH_2Br$

$RCH{=}CH_2 + HCl \rightarrow RCHClCH_3$

$RCH{=}CH_2 + HCl$ (in H_2O) $\rightarrow RCH(OH)CH_3$

$RCH{=}CH_2 + HBr \rightarrow RCHBrCH_3$

$RCH{=}CH_2 + HBr$ (in H_2O) $\rightarrow RCH(OH)CH_3$

$RCH{=}CH_2 + HI \rightarrow RCHICH_3$

$RCH{=}CH_2 + HI$ (in H_2O) $\rightarrow RCH(OH)CH_3$

$$RCH{=}CH_2 + H_2SO_4 \rightarrow R\underset{\displaystyle OSO_3H}{\underset{|}{C}}HCH_3$$

$RCH{=}CH_2 + H_2SO_4$ (in H_2O) $\rightarrow RCH(OH)CH_3$

$RCH{=}CH_2 + H_2SO_4$ (in CH_3OH) $\rightarrow RCH(OCH_3)CH_3$

$RCH{=}CH_2 + Hg(O_2CCH_3)_2$ (in H_2O) $\rightarrow RCH(OH)CH_2{-}HgO_2CCH_3$

$RCH{=}CH_2 + Hg(O_2CCF_3)_2$ (in CH_3OH) $\rightarrow RCH(OCH_3)CH_2{-}HgO_2CCF_3$

WORKSHOP GUIDE

Week 5

Alkenes: Electrophilic Addition Mechanism: Carbocations

This Workshop continues to bridge from ideas about equilibrium (first-year chemistry) to ideas about rate (second-year chemistry). The Workshop starts in *Problem 1* with the ideas of conformational isomers that were central to Workshop 4. The focus, however, is now on conformational change. The interconversion of isomers is treated as a chemical reaction, with emphasis on the reaction energy diagram as a useful tool for describing the changes in structure and energy that correspond to the reaction. The reaction energy diagram leads directly to a discussion about the difference between rates and equilibrium and the structure/energy changes corresponding to these ideas. By using models, students should be able to see how torsional, steric, and angle strain changes as the conformation changes. There will be many good opportunities to connect back to ideas in Workshop 4.

The big challenge in Workshop 5 is to make the connection between the chemical reaction in *Problem 1* (conformational change) and the chemical reaction in *Problem 2* (ethylene and HCl). The **big idea** is that the reaction energy diagram applies to all reactions. Details may differ (for example, there is an intermediate on the path in *Problem 2*), but the general form is the same: there are energy minima for starting material and products, connected by an energy maximum, a barrier. Because ethylene and HCl is a two-step reaction, the reaction energy diagram is just the sum of the simple minimum - maximum - minimum diagrams for each step. It would be good to work out *Problem 2* in this manner; construct the diagram for the individual steps and then "add" them up to make one diagram with a reaction intermediate.

Problem 3 introduces the new idea of competitive reactions. (Remember Workshop 3, *Problem 4*) This could be approached as in *Problem 2*; construct the separate reaction energy diagrams for the two different "first steps" and then superimpose the diagrams to reveal the differences. *Problem 3* is a good example of the need to be clear about the difference between what we know and how we explain what we know. We *know* that 2-bromo-2-methylpropane is the product formed fastest, because that is what is *observed.* Our mechanism and the corresponding reaction energy diagram is a hypothesis that rationalizes the observation.

Problem 4 is intended to address the sense of shock that results from confronting twelve reactions of alkenes. The **big idea** is that our ideas of mechanism give us organizing principles that reveal that all twelve reactions are the same; they are just specific examples of a more general idea. This is a breakthrough idea. We can learn twelve reactions by understanding the general idea; we do not have to memorize twelve unrelated, disconnected sets of magical symbols. There is a better way to learn this material.

WORKSHOP

Week 6

Review

1. **Writing Reaction Mechanisms.** A reaction mechanism is a detailed description of the way a reaction occurs and is based on the known experimental data about the reaction. Some reaction mechanisms are well established. In other cases they may be speculative and are likely to change as more data become available. Mechanisms map the path by which the reactants change into products, and the movement of electrons that accompanies this change. They also show how reactants come together, the intermediates formed, and structures of transition states. Bond cleavage and formation is shown with the help of curved arrows. Movement of one electron is specified with half an arrowhead, and the movement of two electrons is specified with a full arrowhead, as shown in the following examples:

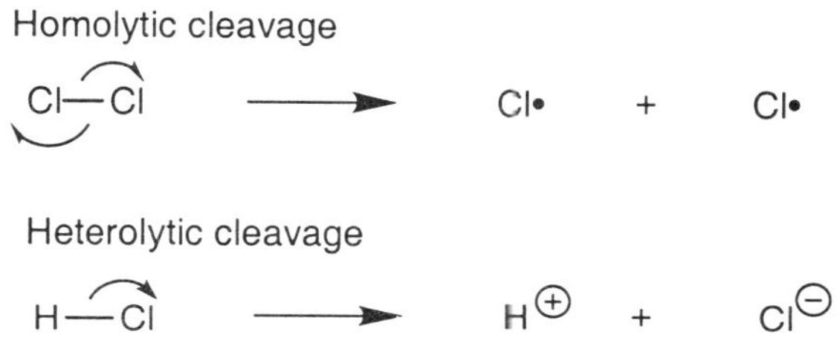

In this Workshop we will consider the reactions of a three-membered cyclic ether, also known as an epoxide, with water to give a 1,2-diol. This ring opening reaction takes place in an aqueous solution under acidic or basic conditions. The ring opening is very slow in water at pH 7. We will work through a series of steps to come to an understanding of possible mechanisms for these reactions.

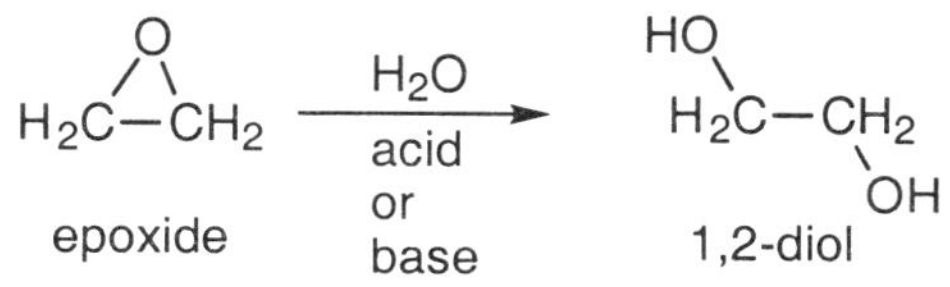

A. First, Consider Basic Conditions

1. Which reactive species is present in basic solution in water?

2. What is the charge on the reactive species? Characterize the reactive species as Lewis acid or a Lewis base.

3. Identify the possible site(s) of reactivity in the epoxide molecule. Characterize these sites as Lewis acids or Lewis bases.

4. With the help of arrows show the movement of electrons that corresponds to reaction between the epoxide molecule and the reactive species present.

5. Write the structure of the first-formed product of the reaction proposed in step 4.

6. Is this intermediate product cyclic or open chain?

7. What will be the charge on this intermediate? Is it a Lewis acid or a Lewis base?

8. Show with the help of arrows how the free diol is formed from this intermediate.

9. Identify all the Lewis acids and bases in the preceding steps.

10. Summarize all the steps in the overall reaction and show structures for all reactants, intermediates, and products. Use arrows to show electron movements.

11. Is this reaction catalytic? Explain.

B. Now, Consider Acidic Conditions

1. Which reactive species are present in an acidic solution in water?

2. What is the charge on the reactive species? Characterize the reactive species as Lewis acids or Lewis bases.

3. Identify the possible site(s) of reactivity in the epoxide molecule. Characterize these sites as Lewis acids or Lewis bases.

4. With the help of arrows show the movement of electrons that corresponds to reaction between the epoxide molecule and the reactive species present.

5. Write the structure of the first formed product of the reaction proposed in step 4.

6. Is this intermediate product cyclic or open chain?

7. What will be the charge on this intermediate? Is it a Lewis acid or a Lewis base? Identify the possible site(s) of reactivity in this intermediate.

8. Identify a species in solution that will react with this intermediate.

9. Show with the help of arrows how the intermediate reacts with a species in solution. (Remember that the net reaction involves ring opening.)

10. Write the structure of the first-formed product of the reaction proposed in step 9.

11. What is the charge on this first-formed product?

12. Show with the help of arrows how the free diol is formed from this first formed product.

13. Identify all the Lewis acids and bases in the preceding steps.

14. Summarize all the steps in this reaction and show all the intermediates. Use arrows to show electron movement.

15. Is this reaction catalytic? Explain.

2. Give a reasonable mechanism (use curved arrows to show movement of electron pairs) for the acid-catalyzed polymerization of 2-methylpropene shown below. You need to show just a few of the polymer chain lengthening steps.

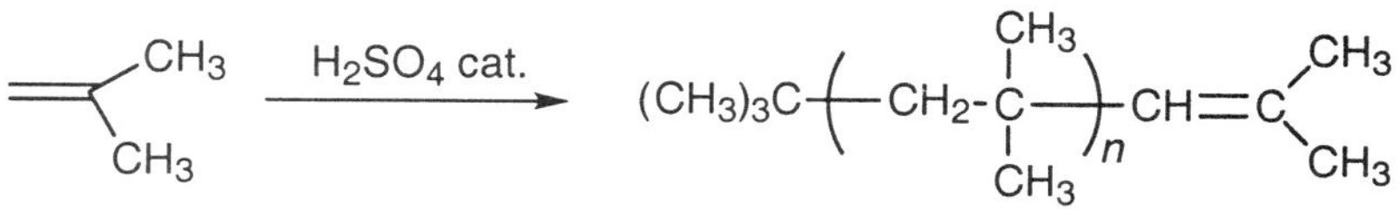

polyisobutylene
n = thousands

3. Consider the data about C-H bonds in the following table. Discuss the variations in bond length, bond dissociation energy, IR stretching frequency, and pK_a in terms of the observed geometries and our ideas relating structure and bonding. Be sure to explain the trends within the categories.

Molecule	∠CCH	Bond Length, Å	BDE in kJ/mol	BDE in kcal/mol	IR Stretch, cm^{-1}	pK_a
Ethane	109.6	1.10	420	100	2850-2960	~60
Ethylene	121.7	1.08	444	106	3020-3100	44
Acetylene	180	1.06	552	132	3300	25

WORKSHOP GUIDE

Week 6

Review

Since Workshops meet every week without fail, it sometimes happens that the class has just had an exam and there is not enough new material for a Workshop. This situation presents a useful opportunity to use a Workshop to consolidate and review. Workshop 6 is an example.

Students are often surprised by the detailed steps that make up a good mechanism. They may think of a mechanism as something to be memorized rather than a hypothetical story that must be consistent with *everything* they know. Workshop 6 gets directly at these two misconceptions and is structured to make the point that a mechanism is a very logical story.

One of the key roles of the Workshop leader is to help the students make connections. Workshop 6 is full of opportunities to do so. Key ideas about acids, bases, ring strain, stoichiometry, Lewis structures, formal charge, stability, and reactivity are embedded in the mechanism.

Although Workshop 6 involves the reactions of an epoxide, the students do not need to know any of the chemistry of ethers or epoxides. The entire Workshop derives from fundamental ideas covered in the first five Workshops. It can be used as an introduction to writing mechanisms or as a review of writing mechanisms (as it is used here). In part, the Workshop makes the point that the fundamental mechanistic ideas *explain* and predict chemistry, that is, a mechanism is both retrospective and prospective.

The second problem continues the idea that a good mechanism developed for one set of observations can be extended to explain (or predict) new observations. Polymerization is often treated as a special topic; it is not. The key chain-lengthening step in acid-catalyzed polymerization is another example of the $E^{\oplus}/N{:}^{\ominus}$ mechanism developed in Workshop 5, *Problem 4*. One of the jobs of the leader is to make sure that this connection gets made. At some stage, the leader might ask the students to identify the electrophile and the nucleophile involved in chain propagation.

Problem 3 is another review opportunity. Conceptual ideas about structure and bonding provide a basis for explaining the trends in the observations as the structure varies from alkane to alkene to alkyne. The first step in working with the problem is to identify clearly what is being measured in each column. Only then does it make sense to analyze the trends.

WORKSHOP

Week 7

Alkenes: Reactions

1. For the following pairs of isomers, circle the *major* species at equilibrium. **Convincingly explain your choices.** (Because this question is about thermodynamics, you do not have to worry about how the reaction occurs or whether it occurs at a significant rate.)

   a. (*Z*)-2-butene $\rightleftharpoons$ (*E*)-2-butene

   b. 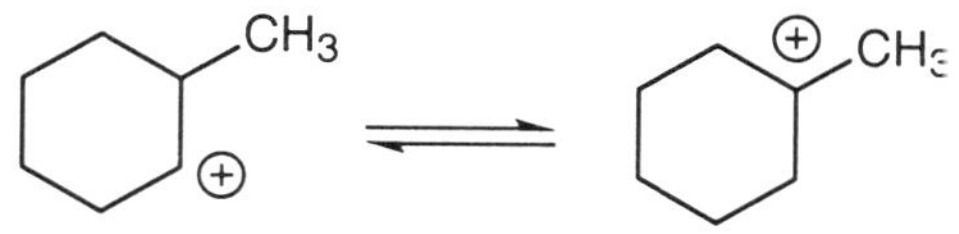

   c. 

   d. $(CH_3)_2CH\text{–}\dot{C}H_2 \rightleftharpoons (CH_3)_3C\cdot$

2. Consider the following mechanistic proposal for the reaction of HBr with alkenes in the presence of light and peroxides.

Initiation: $ROOR \xrightarrow{h\nu} 2RO\cdot$

$RO\cdot + HBr \longrightarrow ROBr + H\cdot$

Propagation 1: $H\cdot + CH_2{=}CH_2 \longrightarrow H\text{–}CH_2CH_2\cdot$

Propagation 2: $H\text{–}CH_2CH_2\cdot + HBr \longrightarrow H\text{–}CH_2CH_2\text{–}Br + H\cdot$ etc.

Evaluate this proposal in terms of what you know about the addition of HBr to unsymmetrical alkenes under free-radical (homolytic) conditions. Make use of everything you know to decide whether this is an acceptable or unacceptable mechanism for the reaction of HBr with alkenes. **Explain your reasoning.**

3. **"What's it good for?"** Propose a simple reaction or pair of reactions that will accomplish the following purposes:

Distinctions: Propose experimental tests that would allow you to distinguish the members of the following pairs of compounds. Explain exactly what you would do and exactly what would you expect to observe.

a. Distinguish cycloalkanes and alkenes.

b. Distinguish between 1-methylcyclohexene and methylene cyclohexane 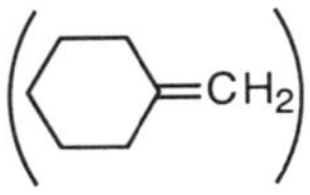.

c. Distinguish an 1-alkene from a 1-alkyne by spectroscopy.

d. Distinguish 1,2–butanediol from 1,3– or 1,4–butanediol.

Transformations: Propose reactions to accomplish the following transformations. Specify reagents, essential conditions, catalysts.

e. Add water to a double bond with two different regiochemistries.

f. Make a ketone.

g. Make 1-bromo-2-methylpropane.

h. Convert a *cis*-alkene to a mixture in which the more stable *trans*-alkene predominates.

4. Compound **J** has the molecular formula C_6H_{12}. Reaction with H_2O/H_2SO_4, or with $BH_3{\cdot}THF$ followed by treatment with a basic solution of hydrogen peroxide, gives the same compound **K**, $C_6H_{14}O$. Reaction of **J** with O_3 followed by reduction with Zn/HOAc, or with $KMnO_4$ in acidic solution, gives only one product **L**, C_3H_6O. Propose structures for compounds **J**, **K**, and **L**.

Observation	**Deduction**

5. **M** and **N** are isomers, C_4H_8. **M** reacts with H_2 in the presence of a catalyst to give an alkane, C_4H_{10}. **N** also reacts with H_2 under the same conditions to give C_4H_{10}, which is different from the compound obtained for **M**. Reaction of **M** with O_3 followed by treatment with Zn/H_3O^+ gives two products, CH_2O and C_3H_6O. **N** reacts under the same ozonolysis conditions to give one product, C_2H_4O. When **N** is heated with a few drops of H_2SO_4, it is converted to a mixture of **N**, **O**, and **P** in which **O** predominates. **N**, **O**, and **P** are all reduced with H_2/PtO_2 to the same alkane. Ozonolysis of **O** gives one product, C_2H_4O, the same product obtained from ozonolysis of **N**. Ozonolysis of **P** gives two products, CH_2O and C_3H_6O. This C_3H_6O compound from **P** is not the same as the C_3H_6O compound from **M**.

Observation	**Deduction**

WORKSHOP GUIDE

Week 7

Alkenes: Reactions

Questions about the relative rates of competing reactions are central to an understanding of organic chemistry. A good mechanism will resolve the issue of "faster or slower." Selective synthesis is based on a step-by-step analysis of relative rates. In order to predict activation energies for competing reactions, we estimate the energies of isomeric transition states. In turn, we estimate the relative energies of isomeric transition states (unstable, energy maxima) by analyzing the relative energies of the stable (energy minima) molecular species that are closest in structure to the transition states. Thus, questions of relative stability (thermodynamics) enter into discussions of relative rate. *Problem 1* emphasizes questions of relative stability. There are different ways to analyze the equilibria. Reference to an experimental result is always acceptable. For example, the acid-catalyzed isomerization of either *Z*-2-butene or *E*-2-butene gives the same Z/E = 24/76, indicating that the Z isomer is less stable than the E form. This type of answer is always correct, and nothing more needs to be said. Alternatively, one can construct a story (a hypothesis) that rationalizes the experimental observation. There is more steric strain in the Z isomer because the van der Waals radii of the methyl groups overlap. That steric strain is absent in the E isomer. Hence the Z isomer is higher energy, less stable, less at equilibrium. Given our desire to explain things, this is also an acceptable answer. Students often fail to recognize that these are different types of answers.

Problem 2 gets right to the heart of an epistemological question about reaction mechanisms. How do we know whether the proposed mechanism is a good one or a bad one? Many students will refer to *external* authority: "That is not what we learned in class." "The book said ..." or "the teacher said ..." Our objective is to teach students to learn to refer to *internal* authority, that is, to observations, to data and to processes of logic. The clue to this problem is in the next-to-last sentence. 1-Bromo-2-methylpropane is the *observed* (known, true) product of the reaction of HBr with isobutene. Does the proposed mechanism successfully predict this product?

Problem 3 examines at reactions of alkenes from a different points of view. The reason that we care about (bother to learn) the reactions is because we can do something with them. Chemists do something: They deduce structures. They make molecules. They solve problems. They do so by putting to work the reactions that they know. *Problems 4* and *5* expand on this **big idea** in the context of structure proof puzzles.

WORKSHOP

Week 8

Alkynes: Synthesis

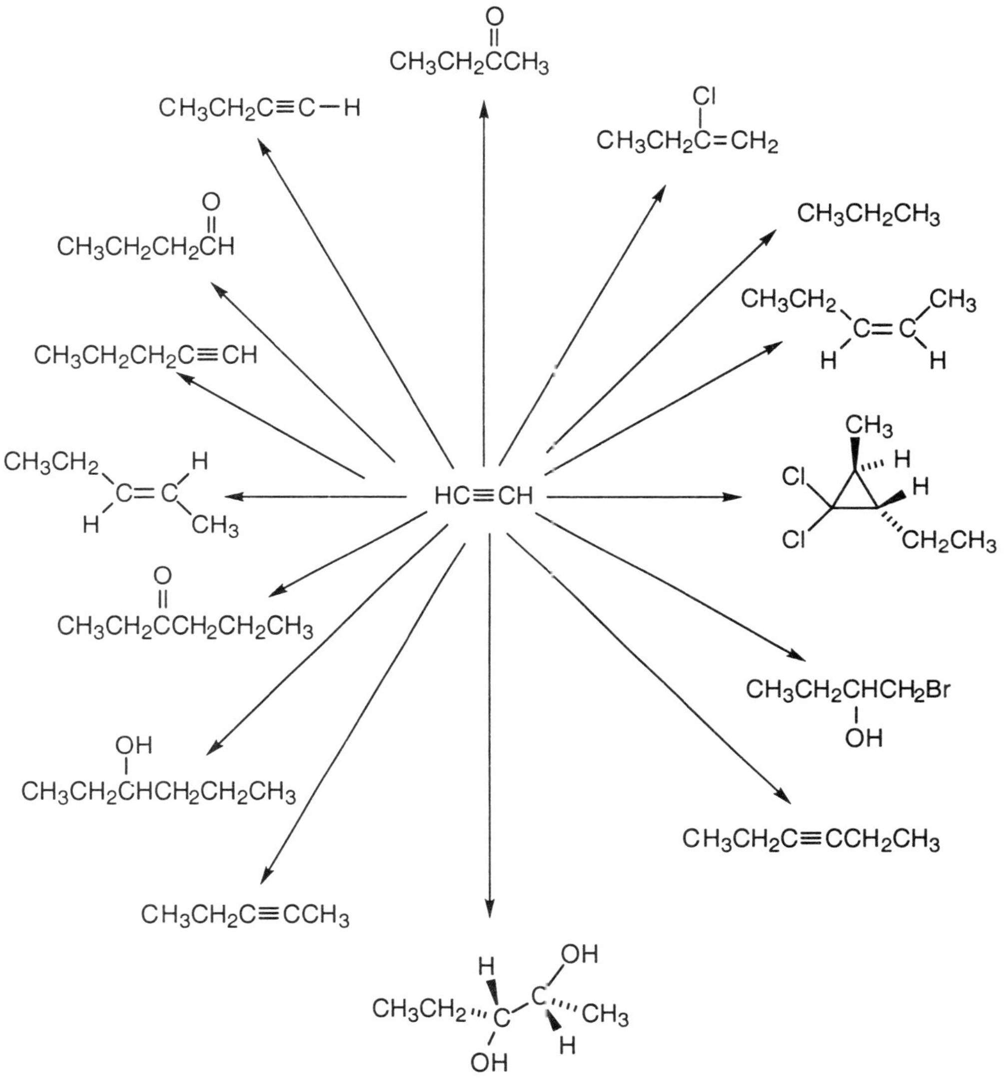

WORKSHOP GUIDE

Week 8

Alkynes: Synthesis

Most of the reactions of alkenes are *functional group interchanges*: one functionality is transformed into another. The chemistry of alkynes provides the first opportunity for making new carbon-carbon bonds and, therefore, the first opportunity to start to think about *synthetic organic chemistry*. All the problems in Workshop 8 involve a bond-making reaction (BMR). Almost all of the problems also require a *functional group interchange* (FGI). It is useful to get students to distinguish the different kinds of steps involved in making larger molecules from smaller pieces. The leader might ask students to label individual steps in a proposed synthesis as FGI or BMR, or to analyze what needs to be accomplished in the synthesis in terms of making a bond (or bonds) and changing a functional group. This helps the students identify which reactions to use to get the job done. Clearly, the reactions of alkenes and alkynes that they are learning can be organized into the two categories, FGI and BMR.

The starburst is much more fun than a list of individual problems. The leader might initiate discussion by asking the group to identify the problems that require only BMR steps.

Not everyone has to do every problem; the leader can parcel those problems out to groups of two or three students. Students can then explain their synthetic proposals to one another and complete part of the star. A blackboard helps; the leader might make a huge star on the board while the students are working. In a second round the students can tackle the problems that require both BMR and FGI. Ultimately, there is great satisfaction in working together to complete what looks like a daunting task. Eager students will make a game out of it, trying to make the star even bigger by adding other syntheses or by working around the periphery to make new connections. The team with the most connections wins!

The starburst format emphasizes the need for **mastery** of the basic organic reactions (starting material, reagents and conditions, products) and an immediate sense of what is accomplished by the reaction. Synthesis is a creative act; it is fun precisely because it is creative. Students who think that the point of organic chemistry is to memorize flash cards will never get to the fun. The point of the organic course is to learn to be synthetically and analytically creative. To speak in sentences, one has to have mastered the vocabulary. The basic organic reactions are the vocabulary of organic chemistry.

WORKSHOP

Week 9

Synthesis

1. Show how to carry out the following synthetic transformations efficiently, specifying reagents and reaction conditions. Each transformation requires at least two steps.

Tactics for conquering synthetic problems:

a. What could be the penultimate (next to last) compound in the synthetic sequence?
b. How could you convert the penultimate compound to the ultimate compound?
c. If you cannot answer (b), then recycle to (a) and try another precursor to the final product.
d. When you get a good final step, then figure out how to make the penultimate compound, following the same logic until you get to the starting material.

i. $CH_3CH(OH)CH_3 \longrightarrow CH_3CH_2CH_2\text{-}Br$

ii. 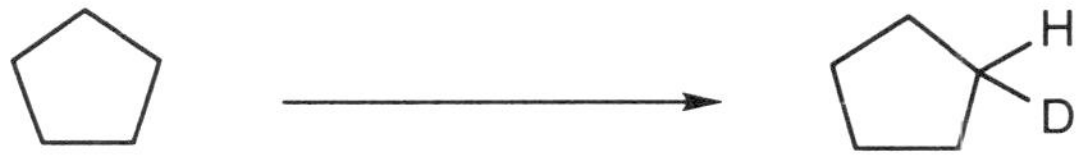

iii. $CH_3CH_2CH_2CH{=}CH_2 \longrightarrow CH_3CH_2CH_2CH_2CH_2\text{-}Cl$

iv. $CH_3CH_2\text{-}C{\equiv}C\text{-}CH_2CH_3 \longrightarrow$ 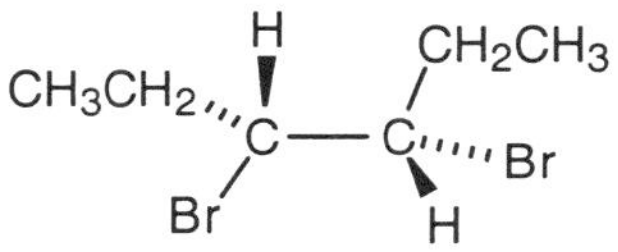

v. $(CH_3)_3CH \longrightarrow (CH_3)_3C\text{–}CO_2H$

vi. $(CH_3)_2CHCH_2CH_2CH_2Cl \longrightarrow (CH_3)_2CHCH_2CH_2CH_2OH$

2. Show how to carry out the following synthetic transformations using any necessary organic and inorganic reagents. More than one step may be required. Indicate with an asterisk all the products that will be formed as racemic mixtures.

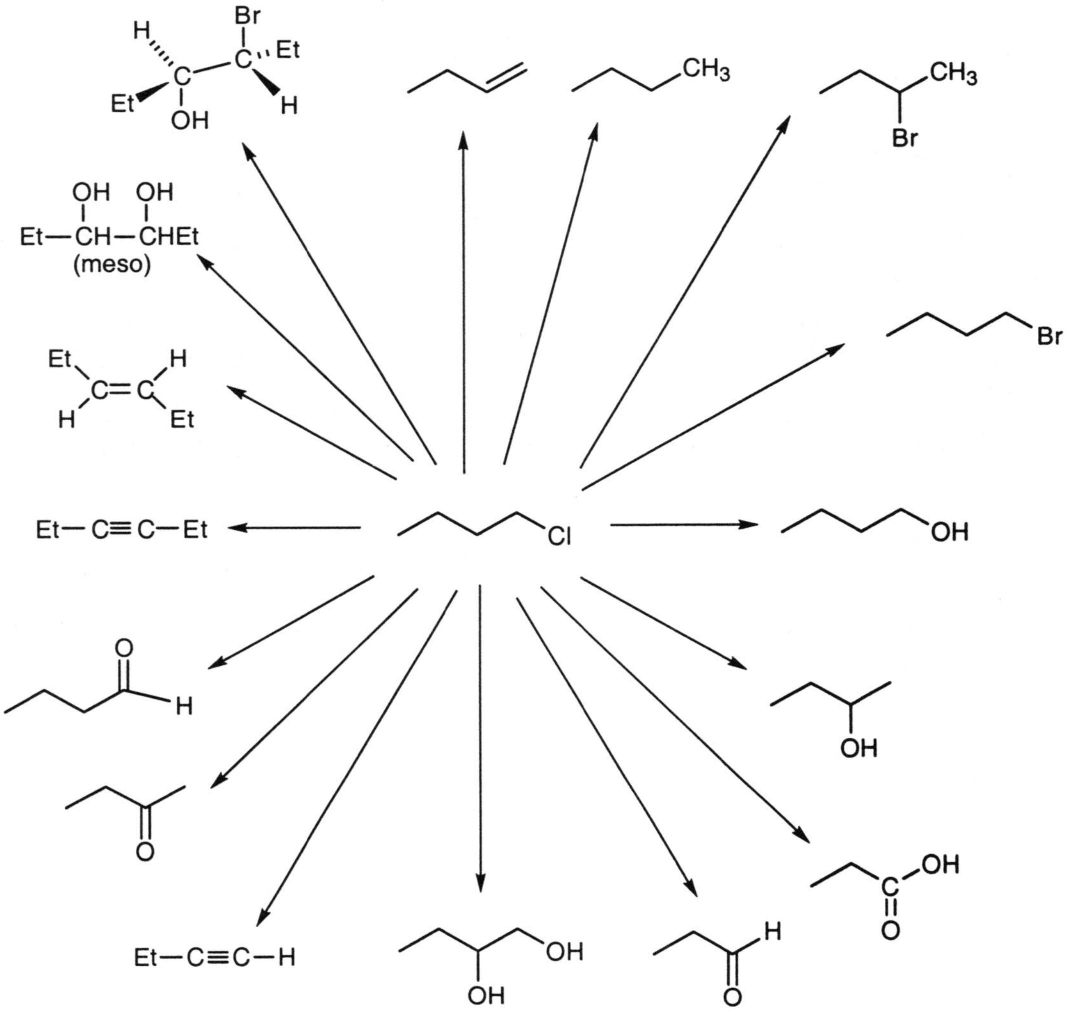

WORKSHOP GUIDE

Week 9

Synthesis

Workshop 9 continues the focus on synthesis begun in Workshop 8. All the ideas in Workshop 8 apply. Workshop 9 makes the retrosynthetic analysis explicit. A good retrosynthetic analysis will be like a tree, with many branches at each junction. The leader might work as scribe and get the group to work on constructing a full tree. *Problem 1(a)* will work well; there are at least two obvious immediate precursors to the product, and, of course, the precursors have precursors. Once the tree is drawn, students can debate the merits of different branches.

Problem 2 extends the synthesis challenge by incorporating new reactions. The leader must have a very keen sense of which reactions are available to the students at the time they do Workshop 9. By the end of the semester there will be many different ways to accomplish the necessary changes. Earlier in the term, the number of variations will be more restricted. Similarly, the identification of racemic mixtures may or may not be appropriate at the time that Workshop 9 is presented. Clearly, a discussion about stereoisomerism must be part of *identifying* the racemic mixtures.

Some leaders have worked this problem by encouraging students to identify the key synthetic intermediates that serve as branch points to several other compounds, for example, 1-butene, 1-butyne, 3-hexyne. This organization of the reactions emphasizes the connections between the reactions and creates a concept map of the transformations. Breaking the problem into smaller pieces also makes it less intimidating.

WORKSHOP

Week 10

Chirality

1. Tell whether the compounds in each of the following pairs are **nonisomeric**, **identical**, **constitutional isomers**, **enantiomers**, or **diastereomers**. Also, assign a **configuration (*R* or *S*)** to all stereocenters. Identify the **optically active (chiral)** compounds. Identify any **meso** compounds. Start by making sure that everyone understands the ideas behind these words. Make molecular models of these compounds to confirm your assignments.

a. 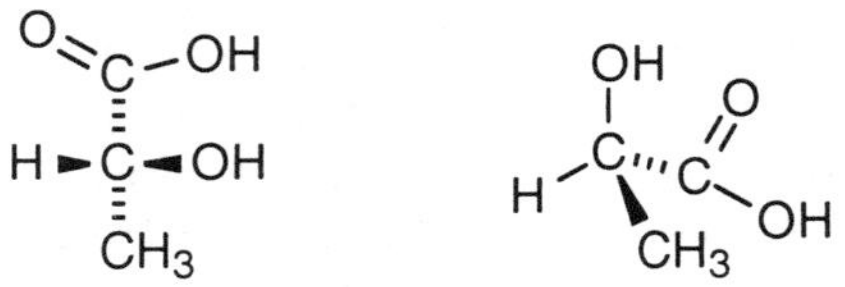

b. 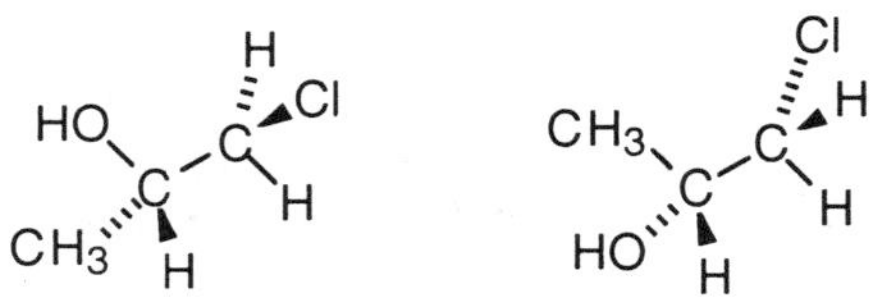

c. 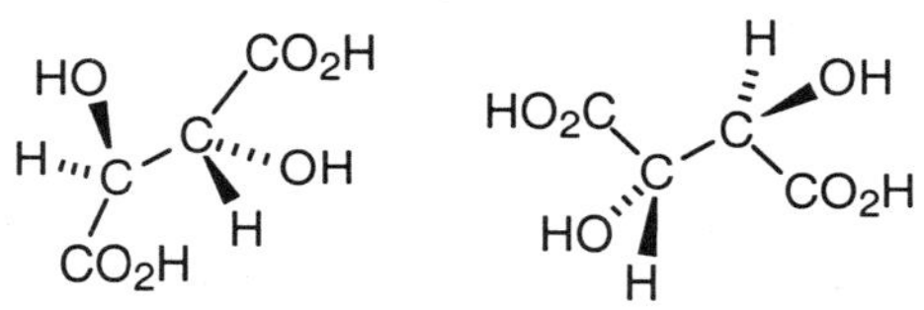

d. 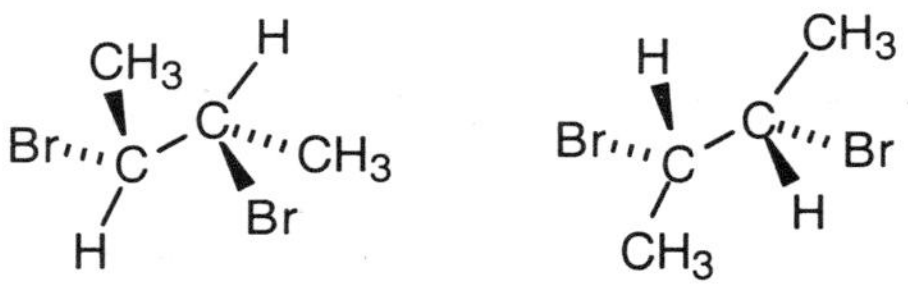

e. 

f. 

2. For each of the following reactions, predict whether the product will be optically active, a racemic mixture, or achiral. **Explain your choice.**

a. (+)–2–chlorobutane $\xrightarrow[\text{in } CCl_4,\ h\nu]{Br_2}$ 2–bromo–2–chlorobutane

b. (+)–2–chlorobutane $\xrightarrow[h\nu]{SO_2Cl_2,\ CCl_4}$ 1,2–dichlorobutane

Several other products are formed.
This one is separated by gas chromatography.

c. (+)–2–chlorobutane $\xrightarrow[\text{EtOH}]{\text{KOH in}}$ $\xrightarrow[\text{ether}]{\text{HBr in}}$ 2–bromobutane

d. *cis*–2–butene $\xrightarrow{Cl_2 \text{ in } CCl_4}$ 2,3–dichlorobutane

e. (+)–1,3–dimethylcyclopentene $\xrightarrow[h\nu]{\text{NBS in } CCl_4}$ 3–bromo–1,3–dimethylcyclopentene

f. (+)-3-methylcyclopentene $\xrightarrow[H_2]{PtO_2}$ methylcyclopentane

g. *cis*–2–hexene $\xrightarrow[\text{2. } NaHSO_3,\ H_2O]{\text{1. } OsO_4}$

3. Peroxycarboxylic acids (RCO_3H) react with alkenes to give three-membered ring cyclic ethers called *epoxides*. For example, 3-hexene reacts with peroxyacids to give 3,4-epoxyhexane.

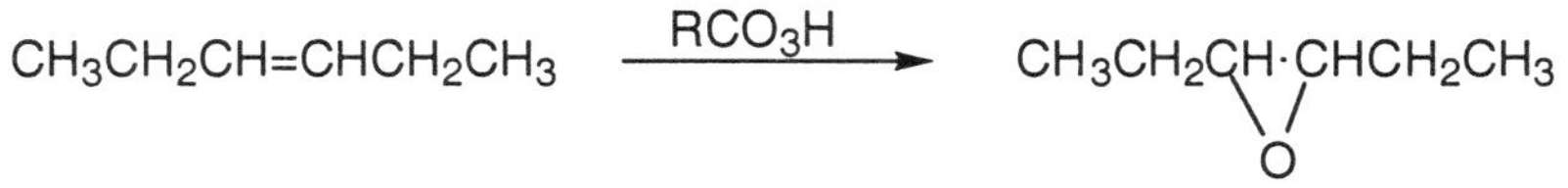

3,4-epoxyhexane

a. This reaction is performed on *cis*-3-hexene. The 3,4-epoxyhexane that is formed **cannot** be separated into optically active molecules.

i. What is the structure of the 3,4-epoxyhexane formed from *cis*-3-hexene?

ii. What is the stereochemical path of the addition of the oxygen to the double bond?
That is, are the bonds to the oxygen formed *syn* or *anti* with respect to the double bond?

b. The 3,4-epoxyhexane from *cis*-3-hexene is reacted with H_2O in the presence of an acid catalyst. The epoxide ring is opened to give a diol, 3,4-dihydroxyhexane.

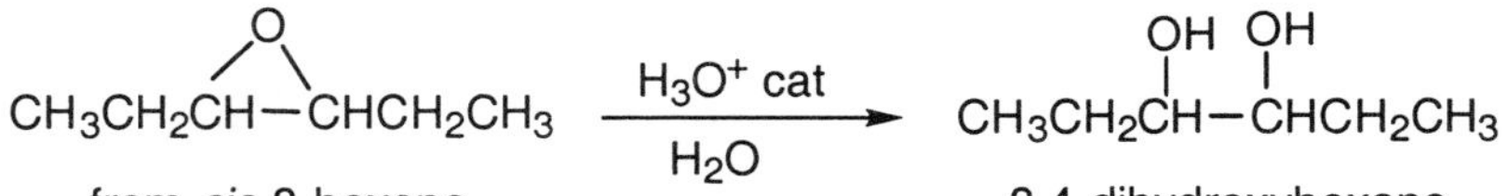

from *cis*-3-hexene   3,4-dihydroxyhexane

This 3,4-dihydroxyhexane **can** be separated into optically active forms, (+)-3,4-dihydroxyhexane and (-)-3,4-dihydroxyhexane.

i. What is the structure of the 3,4-dihydroxyhexane formed from *cis*-3-hexene?

ii. What is the stereochemical path of the addition of water to the epoxide? Is the new C-O bond formed *syn* or *anti* to the C-O bond of the epoxide?

WORKSHOP GUIDE

Week 10

Chirality

Workshop 10 is a particularly good example of the way that a Workshop builds from a simple, definitional knowledge to more sophisticated problem-solving. *Problem 1* clearly calls for a discussion of the precise meanings of the words before work is started on the problem. We all think that we know something, only to find that our understanding is fuzzy. The discussion among the students in a Workshop is a "reality check," without penalty. Imagine a student explaining diastereomers or listening to another student explaining them. Either way, the give and take will sharpen the edges of understanding. Even so, talking is not the same as doing, and the application of the definitions to real molecules will lead to further clarifications. Simultaneously, the problem requires that students have a firm understanding of conventions for representing 3-D molecules on 2-D paper. Students will wrestle with the transition from structures on paper to *R* and *S* designations. There are many devices (steering wheels, hands, clocks) to help students assign *R* and *S*. Before opening a discussion of these visual aids, it is essential to be sure that everyone understands the definitions of *R* and S. The nomenclature system is a convention; one has to know the rules.

Of course, it is reasonable to ask, so what if we can classify and name objects? *Problem 1*, by itself, is sterile. It is the kind of apparently idle exercise that makes students dislike organic chemistry. *Problem 2* moves to the "so what" level. We need to be able to do *Problem 1* with facility so we can figure out how many products we will get when we run real chemical reactions. This same issue came up in *Problem 2* in Workshop 9.

There are other dimensions to *Problem 2*, Workshop 10, as well. As written, several of the questions cannot be answered without a knowledge of the mechanism of the reaction and the stereochemical properties of the reaction intermediates.

Students may also wonder "so what" about mechanisms. A good mechanism provides the basis for prediction. Alternatively, we can reason backward from the nature of the products to the stereochemical characteristics of the reaction (including the mechanism). This logic is worth careful attention by the leader. It is a "two-out-of-three" logic: starting material and mechanism leads to predictions about products, or starting material and products leads to conclusions about mechanism.

Problem 3 continues to explore this two-out-of-three logic, in the context of a reaction that the students do not yet know. The point, of course, is to deduce something about the mechanism of this unknown reaction from the stereochemical characteristics of the starting material and the products. Both *Problems 2* and *3* get at all-important epistemological issues. We know about address mechanism because we can figure it out, not because someone told us it was so.

WORKSHOP

Week 11

Alkyl Halides and Tosylates: Mechanism I

1. a. Iodide ion is a good nucleophile, and sodium iodide is quite soluble in acetone. On the other hand, sodium chloride and sodium bromide have low solubilities in acetone. As a result, the reaction of alkyl bromides and alkyl chlorides with NaI/acetone can serve as a *simple* test reaction, as indicated here:

$$R\text{–}Br + Na^{\oplus}\ I^{\ominus} \xrightarrow{\text{acetone}} R\text{–}I + NaBr \quad \text{(precipitate, solution becomes cloudy)}$$

The reactivity order of the following alkyl bromides with NaI/acetone is

$$CH_3CH_2\text{–}Br > (CH_3)_2CH\text{–}Br > (CH_3)_3C\text{–}CH_2\text{–}Br$$

Write a mechanism for this reaction. Explain how your mechanism accounts for the observed reactivity order. (It will help to build molecular models.)

b. The optical rotation of a solution of (+)–2–bromobutane in diethyl ether does not change with time. However, when tetrabutylammonium bromide is dissolved in this solution, the rotation decreases slowly with time to zero. The NMR spectrum of the solution, however, does not change. Clearly explain what is happening. Construct a graph of the mole fraction of (+)-2-bromobutane and (-)-2-bromobutane as a function of time. Why does the rotation decrease to zero? Why doesn't it become negative?

c. Explain the following observation.

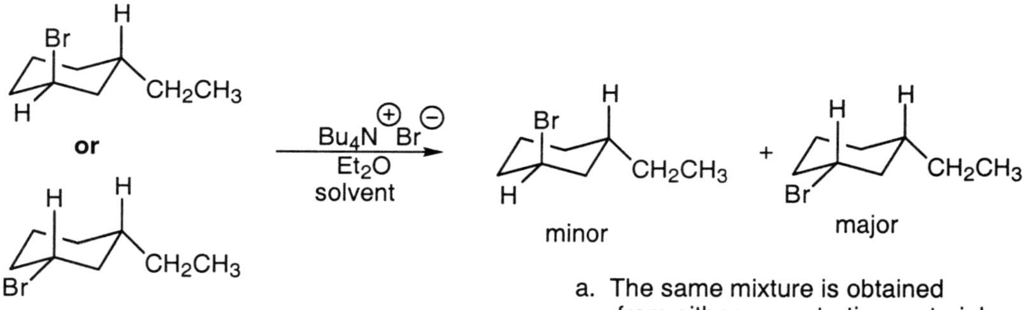

a. The same mixture is obtained from either pure starting material.
b. The rate of isomerization depends on $[Bu_4N^{\oplus}\ Br^{\ominus}]$.

d. Discuss how the three preceding experiments are related to our general understanding of the mechanism of *bimolecular* nucleophilic substitution reactions.

2. a. When 4–chloro–2–methyl–2–pentene reacts with acetic acid (solvent), two substitution products are formed with the rearranged product predominating as shown here. When small amounts of acetate ion are added to the reaction mixture, no increase in rate is observed. Give a mechanism that accounts for these results. Explain clearly why two products are formed.

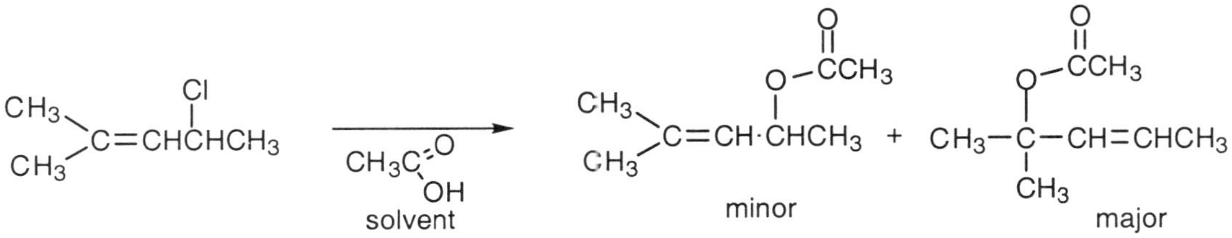

b. Low molecular weight alcohols dissolve in concentrated hydrochloric acid containing $ZnCl_2$. For some alcohols, a reaction takes place to form a new compound that is insoluble in concentrated $HCl/ZnCl_2$. Thus, a clear solution of the alcohol in the reagent turns cloudy. The rate of formation of this new product varies with structure.

$$(CH_3)_3COH > PhCH_2OH \cong CH_2 = CH–CH_2OH \cong (CH_3)_2CHOH > CH_3CH_2CH_2OH$$

What is the insoluble product? Give a balanced equation for the reaction that occurs. Write a mechanism for formation of the insoluble product. Explain how your mechanism accounts for the relative rates of reaction.

c. Discuss how the three preceding experiments (above) are related to our general understanding of the mechanism of *unimolecular* nucleophilic substitution reactions.

WORKSHOP

Week 12

Alkyl Halides and Tosylates: Mechanism II

1. a. Even though the following reaction can potentially give two stereoisomeric elimination products, only one is formed. Give the structure for this alkene.

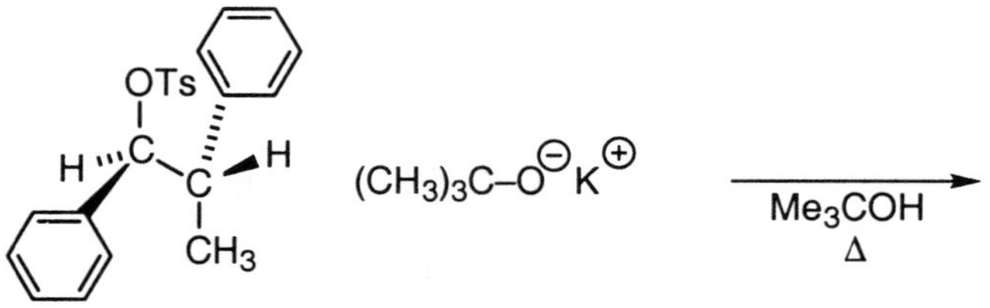

b. The diastereomer of the tosylate shown above also reacts with potassium *t*-butoxide to give an alkene that is diastereomeric to that obtained above. Give the structure for this alkene. Clearly account for the stereochemical course of these two reactions. Why is there no crossover? Explain clearly.

c. Give the products for the following two reactions.

trans–2–methylcyclohexyltosylate $\xrightarrow{Me_3CO^{\ominus} / Me_3COH}$

cis–2–methylcyclohexyltosylate $\xrightarrow{Me_3CO^{\ominus} / Me_3COH}$

d. Discuss how the experiments in parts 1(a), (b) and (c) are related to our general understanding of the mechanism of bimolecular elimination reactions.

2. a. When the following optically active tosylate (**A**) was reacted with CH_3SNa in CH_3OH, the reaction was observed to be second order, and the substitution product shown was formed almost exclusively. Show the mechanism for this reaction, and predict the stereochemistry of the product.

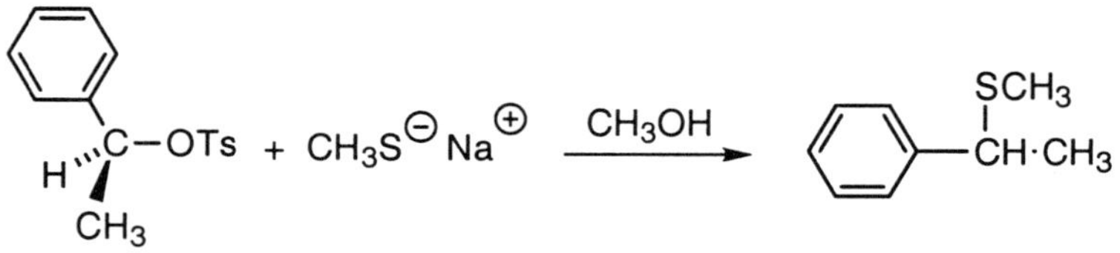

A

b. If the solvent in part (a) was changed from methanol to dimethylformamide (DMF, $HCON(CH_3)_2$), the same product was formed but at a much faster rate. Explain.

c. When **A** was reacted with $NaOCH_3$ in CH_3OH, the reaction was still kinetically second order. However, the major product was the following alkene. Account for the shift from predominant substitution for the reaction in part (a) to predominant elimination for the reaction with CH_3ONa/CH_3OH. Also, show the stereochemistry of the methyl ether substitution product.

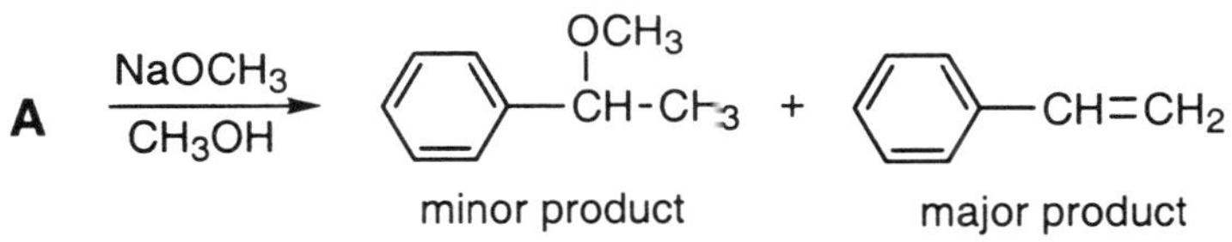

d. How would you change reagents or reaction conditions so that **A** would give almost exclusively the alkene ($C_6H_5CH{=}CH_2$)?

e. When **A** is refluxed in methanol (no added CH_3ONa), both substitution and elimination products are formed, as in part (c). However, substitution predominates, and the stereochemical results are different from those observed for part (c). Explain how and why these results are different from the reaction of **A** with methoxide ion in methanol.

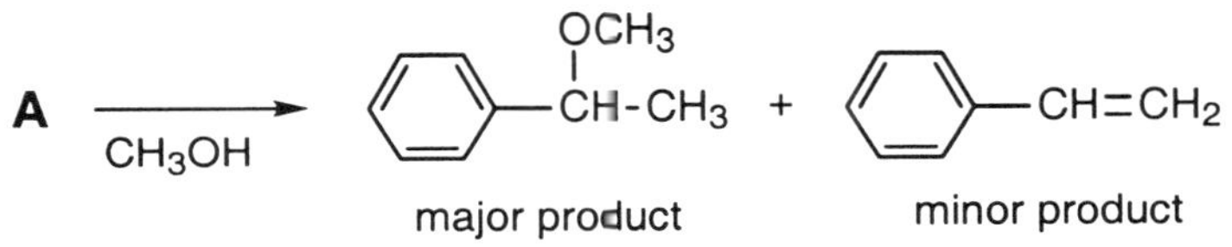

WORKSHOP GUIDE

Weeks 11 and 12

Alkyl Halides and Tosylates: Mechanisms I and II

Mechanisms are deduced from experimental (laboratory) observations. The first three problems in the Workshop emphasize that point. Ultimately, then, *Problem 1* is about epistemology. S_N2 mechanism is the best explanation of the experimental observations in parts 1(a), (b) and (c). Simultaneously, the three problems make use of some fundamental ideas (steric strain, chirality, cyclohexane structure and diastereoisomerism) that were developed in other contexts. It is always good for the leader to make connections: "Remember when we learned about this before."

Problem 2 in Workshop 11 explores the logic that connects observation to deduction (mechanism) for S_N1 reactions.

Workshop 12, *Problem 1*, continues the theme of Workshop 11 for bimolecular elimination reactions.

Problem 2 is a wrap-up problem that turns the logic around and emphasizes how mechanism helps us understand variations in products as a function of variations in reagents and reaction conditions. *Problem* 2 is a good answer to the question: "So who cares about mechanism-what's it good for?" Our mechanisms remove the mystery from the variations in products.

WORKSHOP

Week 13

Alkyl Halides and Alcohols: Synthesis

1. Disconnect the alcohol shown below into three "Grignard-carbonyl" pairs that when combined would give the alcohol.

$$(CH_3)_2\overset{\overset{\textstyle OH}{|}}{C}CH_2CH_3$$

2. Show how to carry out the following synthetic transformations. You may use methanol and ethanol as reactants, any inorganic reagents, and any necessary solvents. If more than one synthetic step is required, show the product for each step.

a. cyclopentane ⟶ deuteriocyclopentane (C_5H_9D)

b. cyclopentene ⟶ 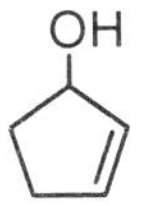

c. 1-chloropentane ⟶ 1-heptanol

d. 2–methyl–1–propanol ⟶ 3–methylbutanenitrile

e. cyclopentanone ⟶ cyclopentyl bromide

f. cyclohexanol ⟶ 1–methylcyclohexene ⟶ *trans*–2–methylcyclohexanol

g. 1–propanol ⟶ 3–pentanone; 1–propanol ⟶ 2–methyl–2–butanol

h. isopropyl bromide (only carbon source) ⟶ $(CH_3)_2CH\overset{\overset{\textstyle O}{\|}}{C}CH_2CH_3$

i. 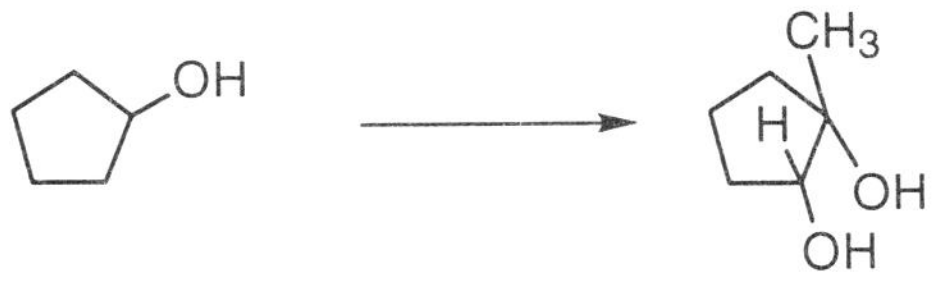

2. Consider the following observations:

a. 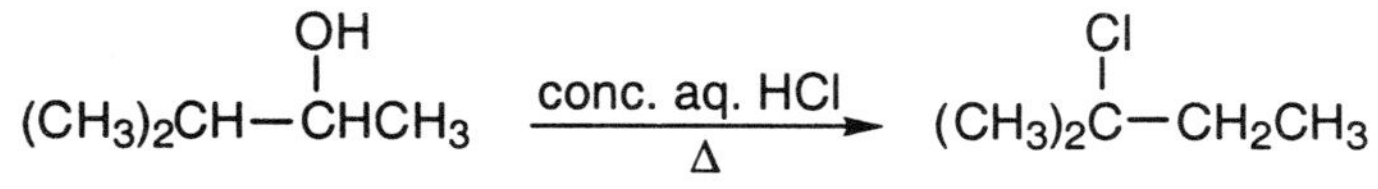

Explain (give a reasonable mechanism) for this reaction

b. When an alcohol reacts with $SOCl_2$/pyridine, the following derivative is formed. If the reactant alcohol is (*S*)-1-deuterioethanol, the product is (*R*)-1-deuterioethyl chloride. Clearly explain how the formation of this intermediate facilitates the conversion of the alcohol to the corresponding alkyl chloride.

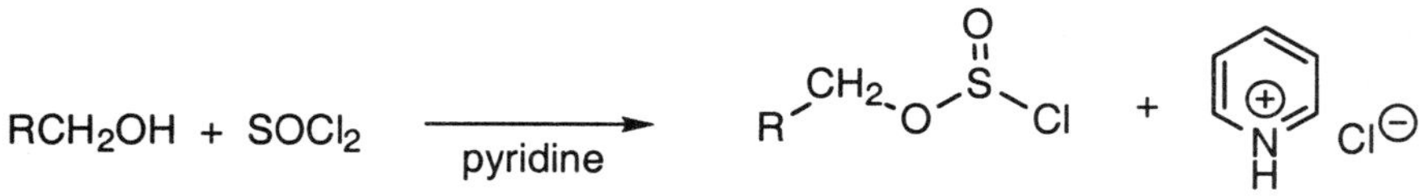

c. The reaction of THF with concentrated HBr gives a new compound, C_4H_9OBr. This new compound shows a broad absorption in the IR at 3400 cm^{-1}. When the new compound is treated with aqueous NaOH solution, THF is re-formed. Propose a structure for C_4H_9OBr, and explain (give a mechanism) for the interconversion.

d. Compounds **A** and **B** both have a molecular formula of C_7H_7Br. When either is reacted with Mg/Et_2O followed by treatment with H_2O, toluene ($C_6H_5CH_3$) is obtained. When **A** is added to $AgNO_3$/EtOH, an immediate precipitate occurs. However, when **B** is added to $AgNO_3$/EtOH, even heating for extended periods of time yields no precipitate. Give structures for **A** and **B**, and explain the reasoning behind your assignments.

Observation	**Deduction**

WORKSHOP GUIDE

Week 13

Alkyl Halides and Alcohols: Synthesis

All the synthesis problems will be made easier by working in a retrosynthetic mode, that is, by disconnecting the product to reasonable precursors, then disconnecting the precursors, and so forth. *Problem 1* is a warm-up problem to get students in the disconnect-Grignard mode before they tackle the syntheses.

We care about the reactions of nucleophiles with alkyl halides and tosylates because they reveal the fundamental pattern of structure/reactivity relationships that we call S_N2. But we also care about these, and related reactions, because we can use the reactions to make new molecules. Workshop 13, *Problem 2* is a reminder that mechanism and synthesis are mutually dependent.

As the synthesis problems get more complicated the teacher and the Workshop leader need to be in tune with what the students have learned and what they have not learned. Many of the problems in this Workshop require Grignard reactions and the redox interconversion of alcohols and carbonyl compounds.

Synthesis and problem-solving are creative acts. The creativity is grounded in mastery of fundamental ideas about structure, reactivity, and mechanism. *Problem 3* is about using the fundamentals to solve interesting puzzles about structure and reactivity.

WORKSHOP

Week 14

Review

1. The following reactions occur by analogous mechanisms. Explain how they are closely related. Also, identify the electrophile and nucleophile for each mechanistic step.

a.

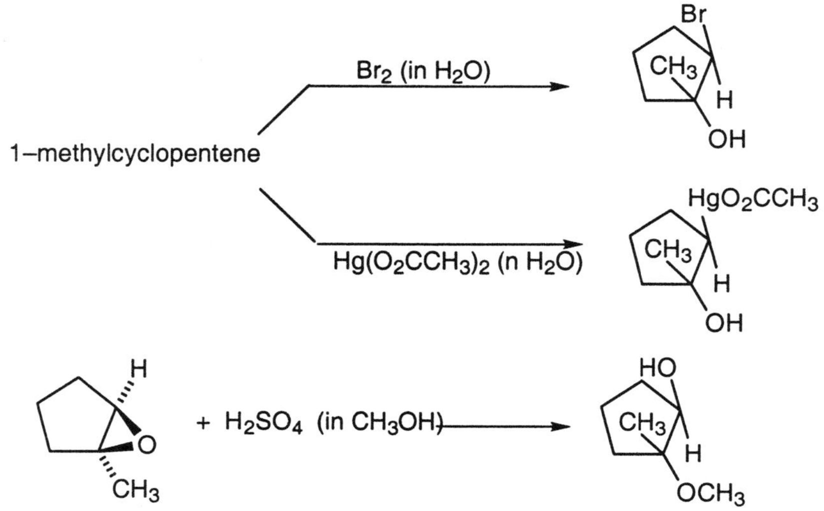

b.

di–isopropyl ether $\xrightarrow[\Delta]{HBr}$ 2 isopropylbromide + H_2O

isopropanol $\xrightarrow[\Delta]{HBr}$ isopropylbromide + H_2O

propene $\xrightarrow{HBr}$ isopropylbromide

c.

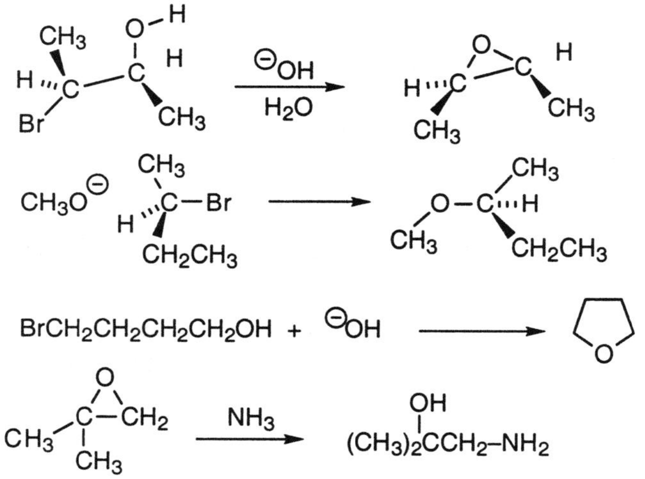

2. Show how to carry out the synthetic conversions indicated in an efficient manner. Find as many other interconversions as you can. If more than one step is required, show the products for each step.

a.

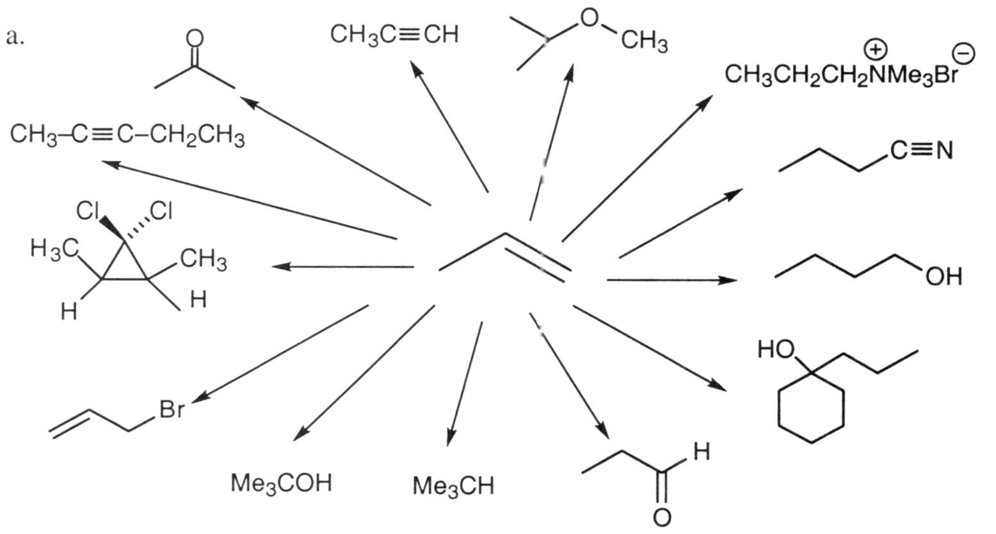

b.

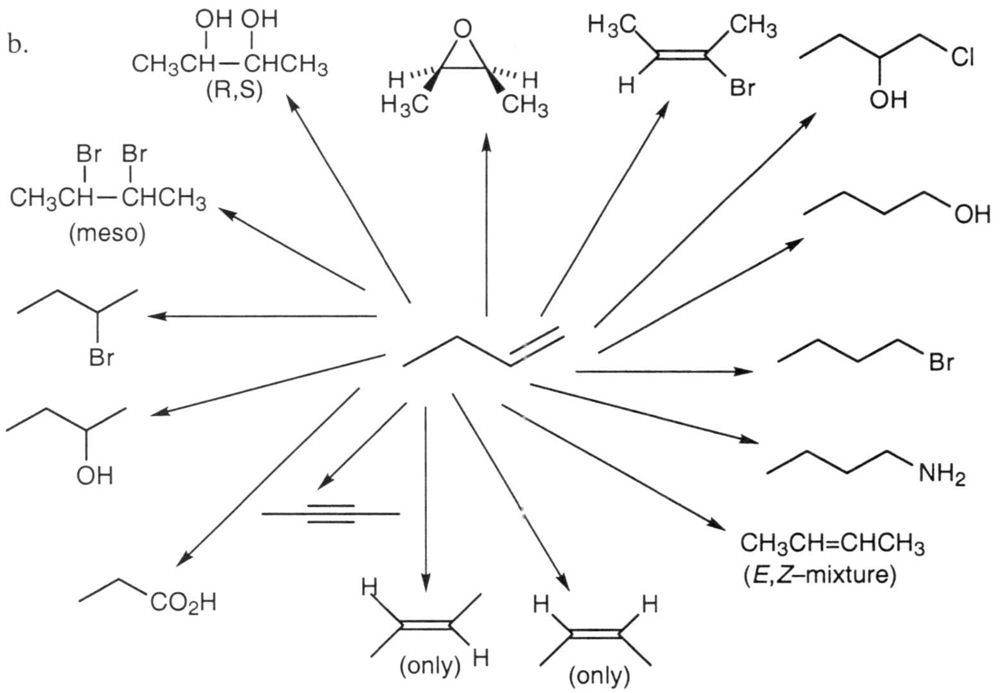

3. For each of the following reaction pairs, tell which would occur at a faster rate. Clearly justify your answer. (*Hint*: what is the mechanism?)

a. 2–butyne + Br_2 $\xrightarrow{\text{in ether}}$ *trans*–2,3–dibromo–2–butene

trans–2–butene + Br_2 $\xrightarrow{\text{in ether}}$ *meso*–2,3–dibromobutane

b.

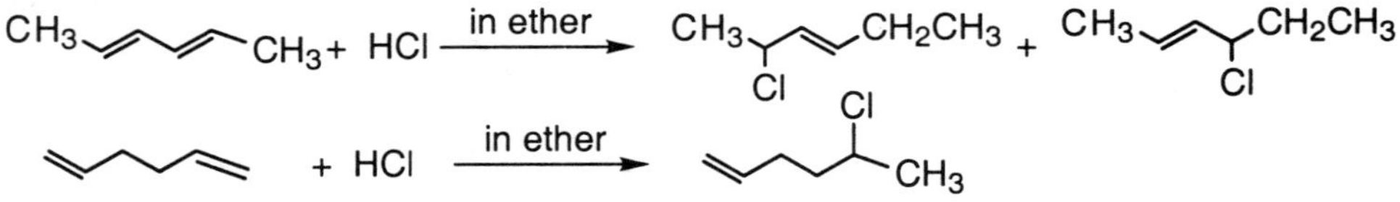

c. 2–chloropentane + $Na^{\oplus} I^{\ominus}$ $\xrightarrow{\text{in acetone}}$ NaCl↓

2–methyl–2–chlorobutane + $Na^{\oplus} I^{\ominus}$ $\xrightarrow{\text{in acetone}}$ NaCl↓

d.

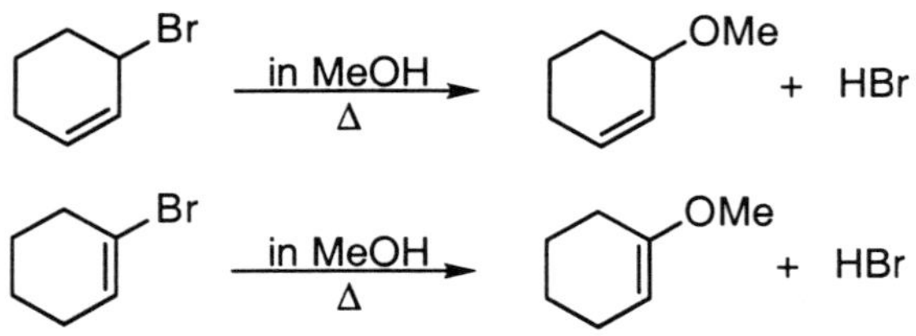

e. *trans*–2–methylcyclohexyl chloride $\xrightarrow[(CH_3)_3COH]{(CH_3)_3CO^{\ominus}K^{\oplus}}$ 1–methylcyclohexene + 3–methylcyclohexene

f. *t*–butanol $\xrightarrow[ZnCl_2]{\text{conc. HCl}}$ *t*–butyl chloride

2–methyl–1–propanol $\xrightarrow[ZnCl_2]{\text{conc. HCl}}$ *t*–butyl chloride

4. For each of the following interconversions, circle the compounds that would predominate at equilibrium. Clearly justify your choice.

a. $CH_3CH_2CH_2{-}MgBr + CH_3C{\equiv}CH \rightleftharpoons CH_3CH_2CH_3 + CH_3C{\equiv}C{-}MgBr$

b. $CH_3CH{=}C(OH)CH_3 \rightleftharpoons CH_3CH_2C(=O)CH_3$

c. $HC{\equiv}CH + {}^{\ominus}OH \rightleftharpoons HC{\equiv}C^{\ominus} + H_2O$

d. $CH_3\overset{\oplus}{O}H_2 + CH_3CH_2NH_2 \rightleftharpoons CH_3OH + CH_3CH_2\overset{\oplus}{N}H_3$

e. 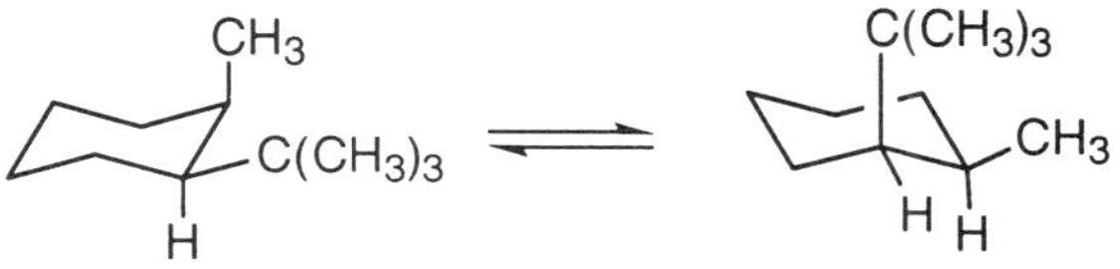

f. 2,3–dimethyl–2–butene $\underset{}{\overset{H_2SO_4}{\rightleftharpoons}}$ 2,3–dimethyl–1–butene

5. Make lists showing as many examples as you know of reactions that

a. form new carbon-carbon bonds;

b. have a specific stereochemistry;

c. involve free-radical intermediates;

d. involve carbocation intermediates;

e. involve oxidation of an organic molecule;

f. involve reduction of an organic molecule.

6. When you arrive at your Workshop, you are presented with a sample of a clear, colorless liquid. Your Workshop leader provides the following information: (1) the sample is pure, and (2) the material belongs to one of the classes of compounds whose chemistry we have discussed in detail in this semester.

Prepare a chart of **Observations** and **Deductions** showing how you would proceed to identify this material. Tell exactly what you would do and what you would observe (**Observation**) and exactly what you would conclude from the observation (**Deduction**). Your Workshop leader will be the "experiment"; that is, when you have specified a procedure and possible observations, the leader will tell you what happened.

Observation	**Deduction**

7. For the following pairs of compounds, suggest a simple, fast, qualitative laboratory procedure that would distinguish the members of the pairs. You should consider both chemical and spectroscopic methods in each case. Be very specific about what you would do and what you expect to observe.

a. 1-hexanol and *n*-pentyl methyl ether

b. 1-hexyne and 2-hexyne

c. *n*-butanol and *t*-butanol

d. 1,4-pentadiene and 1,3-pentadiene

e. 2-bromocyclohexene and 3-bromocyclohexene

f. 2,3-dimethylbutene-2 and 2,3-dimethylbutene-1

g. 2,3-dimethylbutene-2 and cyclohexane

h. 3-bromotoluene and benzyl bromide

i. cyclohexanol and 3-hydroxyhexene-1

j. 1-bromopentane and 3-bromopentane

WORKSHOP GUIDE

Week 14

Review

Studying for a final exam can be numbing if the strategy is simply to go over (i.e., repeat) the semester's work. A much better approach is to find new ways to organize the material and to make new connections. New kinds of problems or, at a minimum, new examples of old problems enliven the process of reflecting and consolidating. Workshop 14 illustrates these points.

The first problem emphasizes a recurring theme. Reactions are learned more easily if they are grouped together by mechanism. *Problem 1* utilizes this strategy for epoxides, emphasizing that the reactions are simply new examples of types of reactions learned previously. The problem also connects back to epoxide chemistry in Workshops 6 and 9.

Faster-slower questions (*Problem 3*) are at the heart of the first-semester discussion about structure and reactivity. As the problem suggests, our mechanistic understanding was constructed to rationalize these faster-slower observations. Structure-stability questions are central to an understanding of equilibrium processes. *Problems 3* and *4* serve as reminders that rate and equilibrium constants are related to distinct energy differences on the reaction energy diagram.

Problem 5 provides a framework for review that requires the student to organize the material in new ways. New connections follow directly from the new organization. If the traditional course is organized by functional group, this problem suggests synthetic and mechanistic ways to organize.

Problems 6 and *7* connect lecture to laboratory. Both turn the traditional **Observation-Deduction** problem inside-out by asking the students to supply the **Observations**. It is easy to imagine working with these problems in an open groupwide discussion, with the leader functioning as scribe.

Students will have the most fun and be most productive if they construct their own problems. All the problems in Workshop 14 can be extended in new directions by students or leaders. Part of the satisfaction in the starburst problems is to try to work around the perimeter or to add new reactions to extend them further and further. In doing so, the students' mastery becomes satisfyingly visible.

Finally, one of the important characteristics of any review is that it identifies areas that need further study. The same value is at work when students review their exam papers or exams from previous years. The point of these useful activities is not to learn answers but to identify gaps in understanding.

WORKSHOP

Week 15

Dienes: Allylic Systems; Diels-Alder

1. Give products for the following reactions. Use words and structures to explain how they are formed.

a.

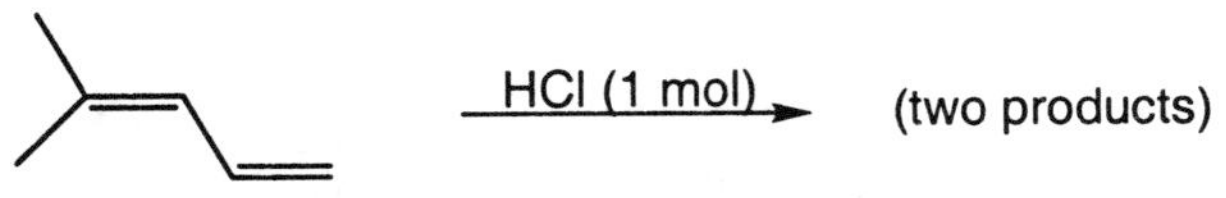

b.

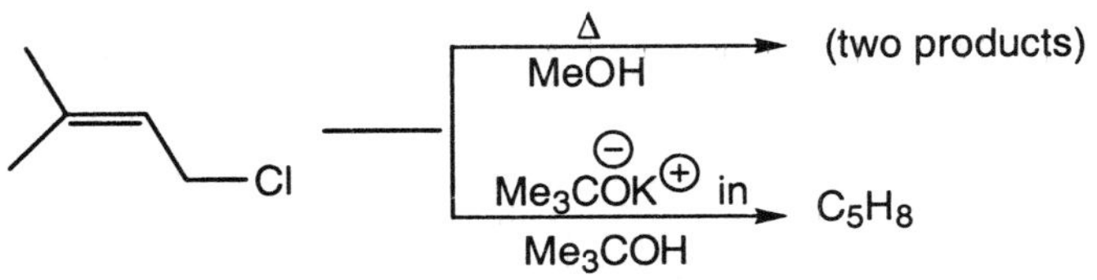

2. For each of the following reactions give reasonable mechanisms that clearly account for the resultant products. Show all important intermediates and resonance structures, and show movement of electron pairs with curved arrows.

a.

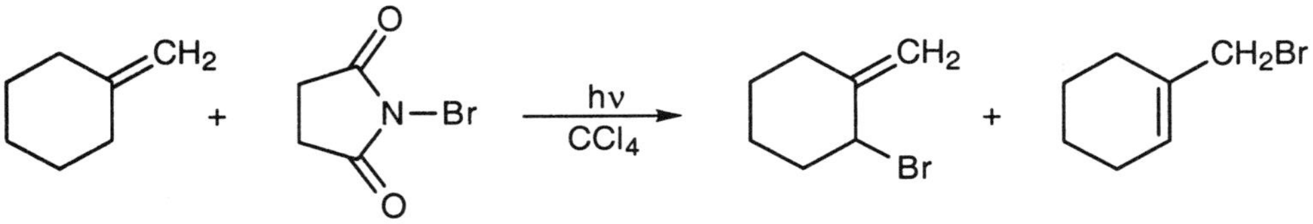

b. $CH_3CH{=}CHCH_2OH$ or $CH_3CH(OH)CH{=}CH_2$ $\xrightarrow[0\ ^\circ C]{HBr}$ $CH_3CH{=}CHCH_2Br$ (75%) + $CH_3CH(Br)CH{=}CH_2$ (25%)

3. Tell which of the following reactions has the *greater* or *greatest* value of the specified property. Clearly explain the basis of your choice.

a.

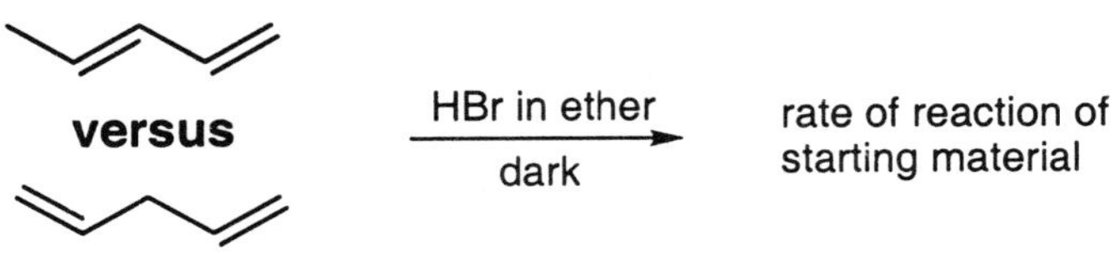

b. 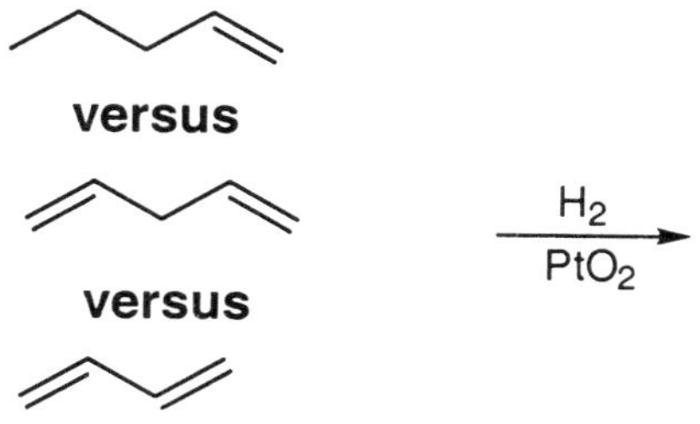

heat of hydrogenation
per double bond

c. 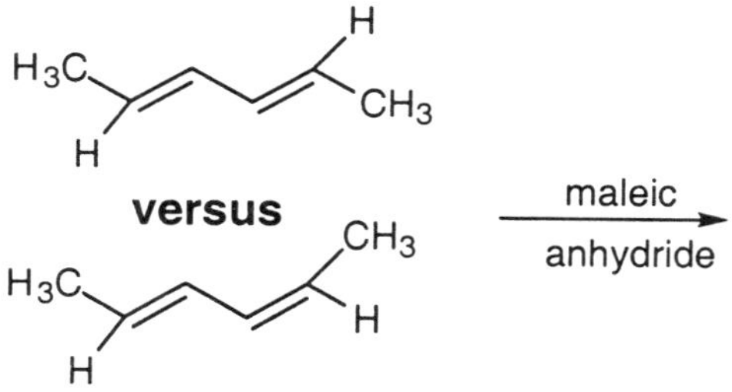

rate of formation
of product

4. Specify the missing reactants or major products in each of the following reactions. Be very careful to specify stereochemistry.

a. 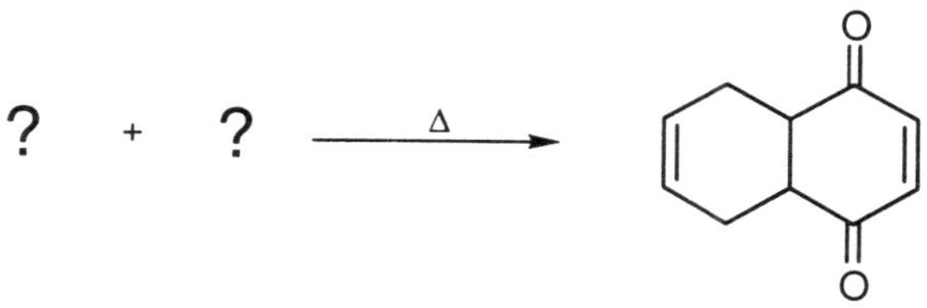

b. 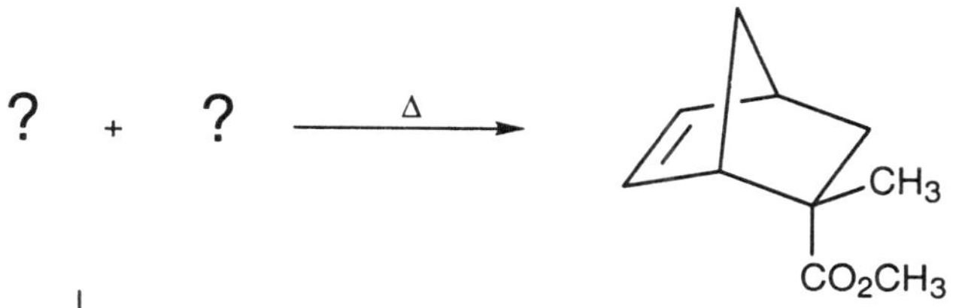

c. 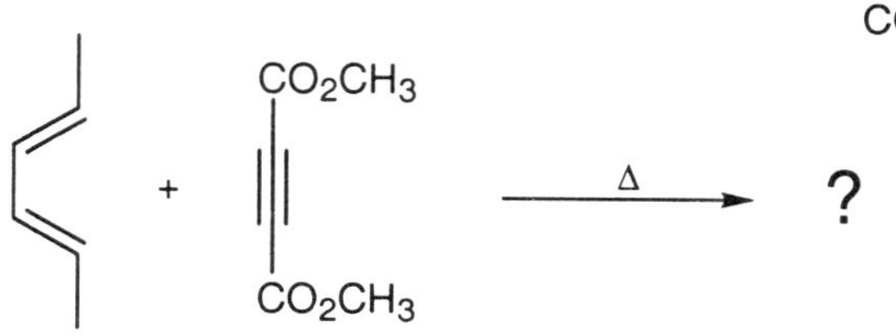

d. 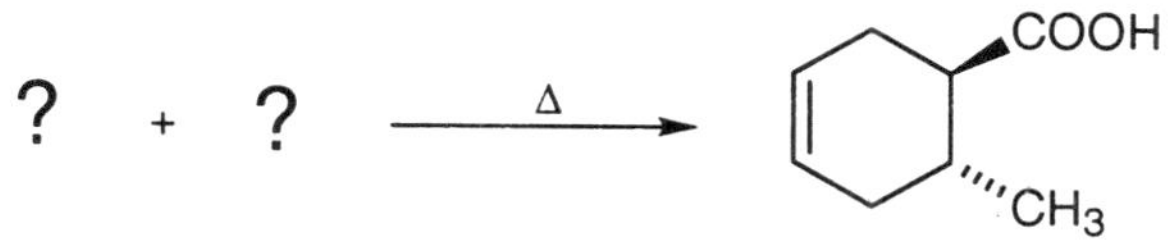

5. It has been reported (Cooley and Williams, *J. Chem. Ed.* **1997**, *74*, 582-585) that the reaction of furan with *N*-phenylmaleimide (**A**) gives a mixture of adducts *exo*-**B** and *endo*-**B**. As indicated below, at a shorter reaction time and lower temperature, *endo*-**B** is the predominant product. At longer reaction times and higher temperatures, *exo*-**B** is the major product. Account for the variation in product ratios. (*Hint*: Consider the reaction mechanism.)

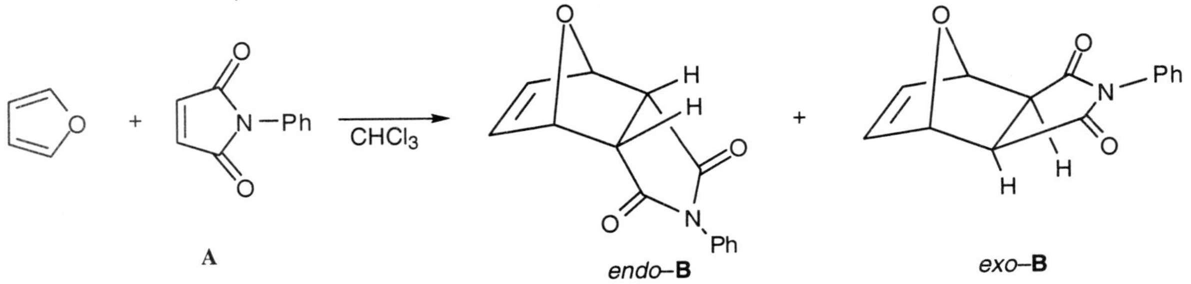

Conditions	A	exo–B	endo–B
7 days at 0 °C	14%	36%	49%
20 days at 25 °C	9%	68%	23%

WORKSHOP GUIDE

Week 15

Dienes: Allylic Systems; Diels-Alder

We learn by making connections. Dienes are especially important in this regard because the chemistry connects back to familiar reactions and mechanisms of reactions of alkenes and alkyl halides. At the same time, dienes connect forward to reactions of conjugated aromatic systems.

The allylic radical, cation, and anion bridge between and from monofunctional compounds to polyfunctional conjugated systems. A good leader will help the students make the connections and build the bridges.

The Diels-Alder reaction occupies a special place in synthetic chemistry because two bonds are made in one reaction, with good control of stereochemistry. *Problem 4* would work well in a round-robin format. *Problem 5* does double duty; it highlights the reversibility of the Diels-Alder reaction and, simultaneously, provides a good example of the distinction between kinetic and thermodynamic control. Beginning students want to argue that the fastest reaction is the one that gives the most stable product (or intermediate). *Problem 5* reveals the fallacy. It is instructive to construct a reaction energy diagram for this problem.

WORKSHOP

Week 16

Aromaticity

1. Predict whether each of the following systems is aromatic, antiaromatic, or nonaromatic. Assume that each system is planar, and justify your choice. Predict specific chemical or physical properties that would reveal the aromatic, antiaromatic, or nonaromatic character of the compound.

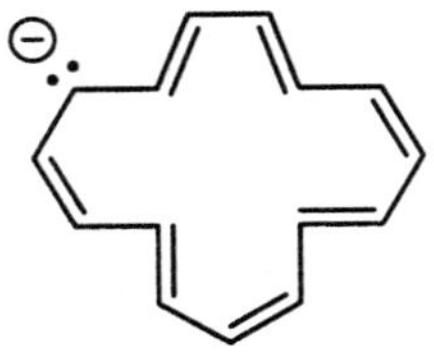

A

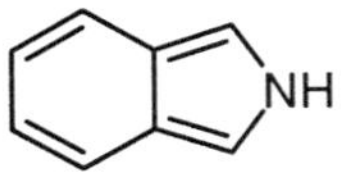

B

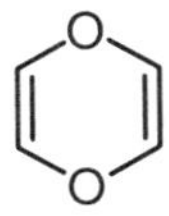

C

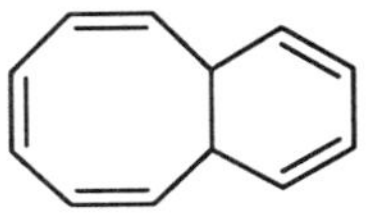

D

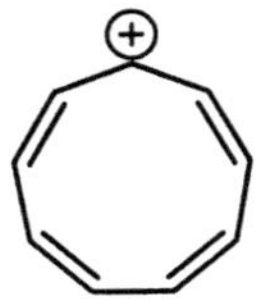

E

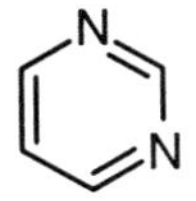

F

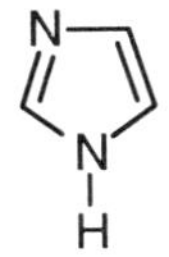

G

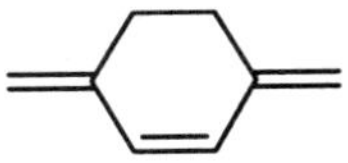

H

2. Provide reasonable explanations for the following observations:

   a. Neither of the following [10] annulenes shows any special stability. (It may help to make models.)

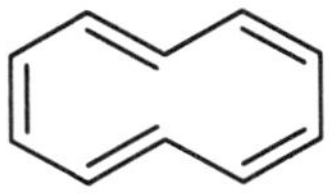

   b. The following [18] annulene gives ^{1}H NMR signals at δ 3.0 and δ 9.3.

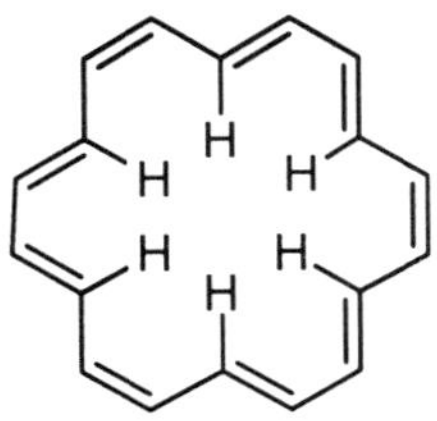

c. The compound shown reacts with HCl to give a weak acid (pK_a ~ 5-6). What is the structure of that weak acid?

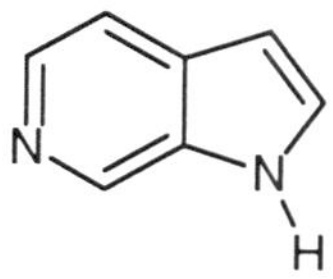

d. Conductivity measurements show that the dibromide shown ionizes to give a dication. Sketch the ^{1}H NMR of the starting dibromide and the dication. Predict the chemical shifts and explain your predictions.

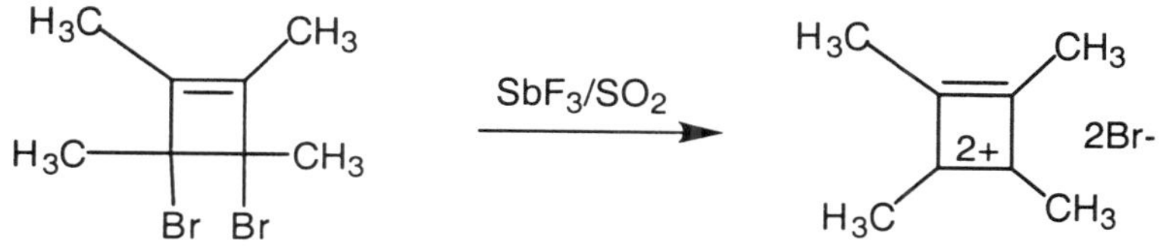

e. Cyclopropenium ion is quite stable, but cyclopropene is not. Diphenylcyclopropenone has a dipole moment equal to 5.08 D, whereas benzophenone (diphenyl ketone) has a dipole moment equal to 2.97 D.

f. The following synthetic procedure was a total failure; the starting material was recovered unreacted.

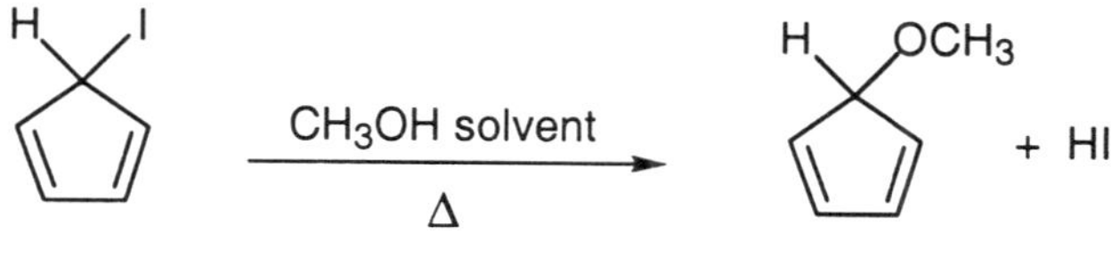

g. 1,3-Cyclopentadiene is extremely acidic for a hydrocarbon (pK_a 15) whereas 1,3,5-cycloheptatriene is extremely nonacidic (pK_a too high to measure).

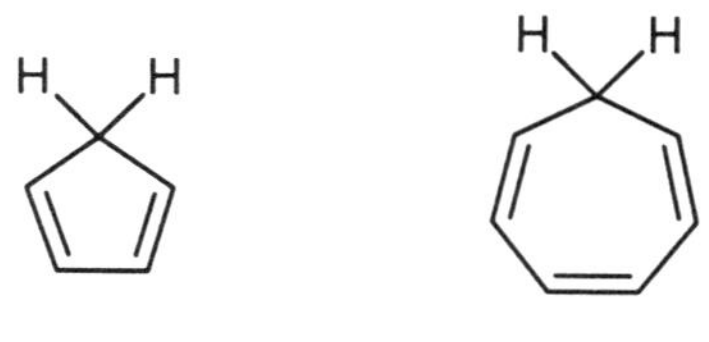

3. Compound **A**, when treated with aqueous sulfuric acid, was converted to isomeric compound **B** (C_9H_{12}). When **B** was heated for an extended time with chromic acid, dicarboxylic acid **C** was obtained. Give the structure of **B** and a mechanism that accounts for its formation from **A**.

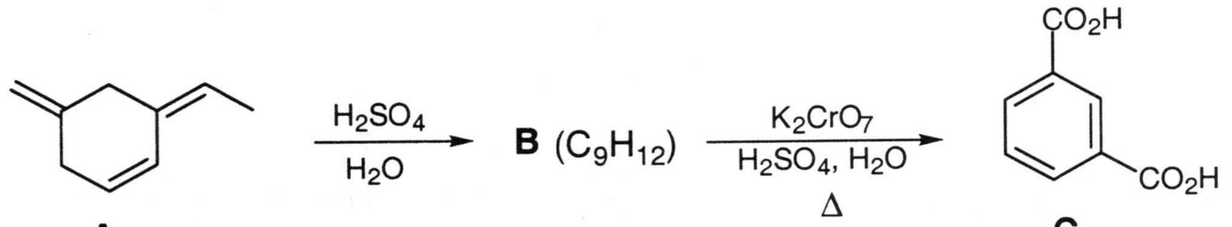

A **C**

WORKSHOP GUIDE

Week 16

Aromaticity

Introductory organic chemistry is about the interplay of structure and reactivity. This connection is at the center of an understanding of the properties and chemistry of cyclic polyunsaturated compounds. The special properties of dienes and allylic systems (Workshop 15) provided some introduction to the idea that special properties are the result of special structures. Workshop 16 explores this **big idea** in a variety of ways.

Problem 1 would be a trivial problem if it did not ask about the chemical *implications* of the results of the counting. The problem will work well in round-robin format; the discussion should bring out the sets of properties of aromatic, non-aromatic, and anti-aromatic systems.

All the other problems in this Workshop explore the connection between structure and chemical and physical properties in the context of ideas about aromaticity. Those connections are the take-home points. One good way to wrap up the Workshop would be to construct a concept map. The three concepts could be aromatic, nonaromatic and antiaromatic. The students could identify the structural, chemical, and physical properties that distinguish and characterize these three concepts.

WORKSHOP

Week 17

Aromatic Electrophilic Substitution

1. Each of the following is classified as a reaction that occurs by an electrophilic aromatic substitution mechanism. Construct a table showing the electrophile and the electrophilic substitution intermediate for each of the reactions. Write a general two-step mechanistic scheme (use curved arrows to show movement of electron pairs) for these reactions.

a. 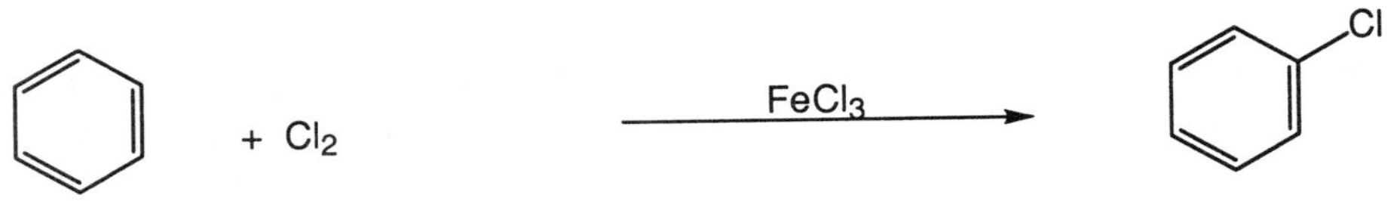

b. 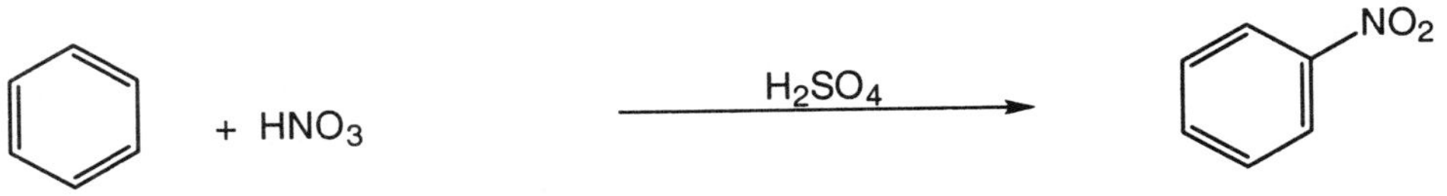

c. 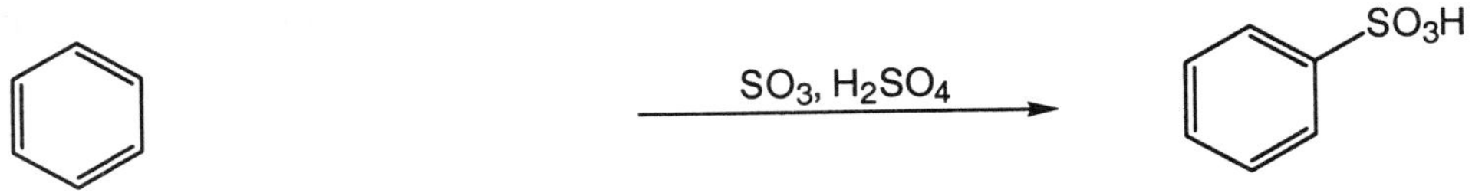

d. 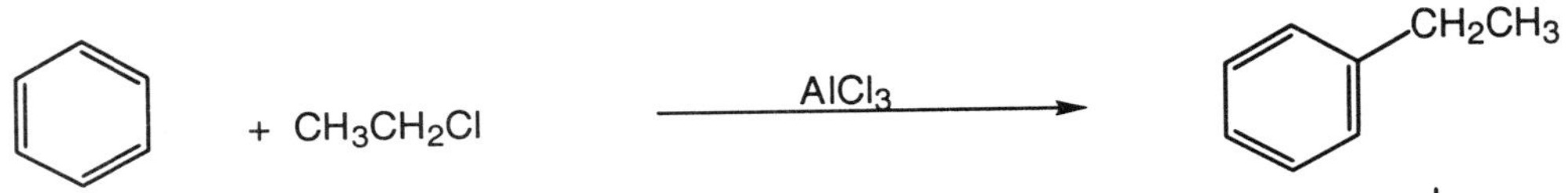

e. 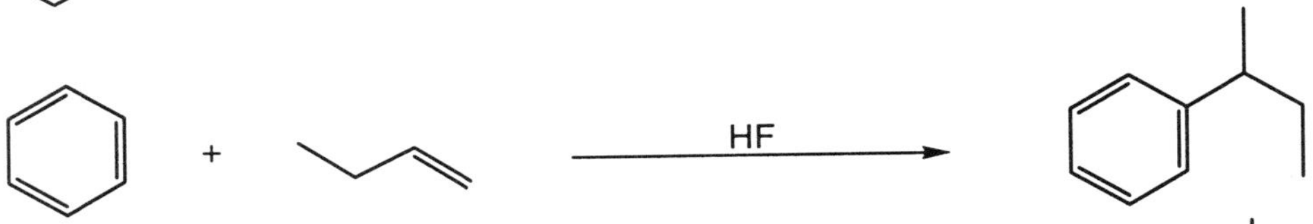

f. 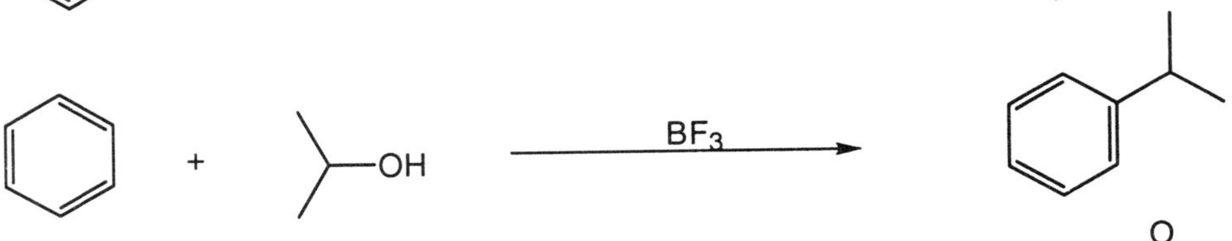

g. 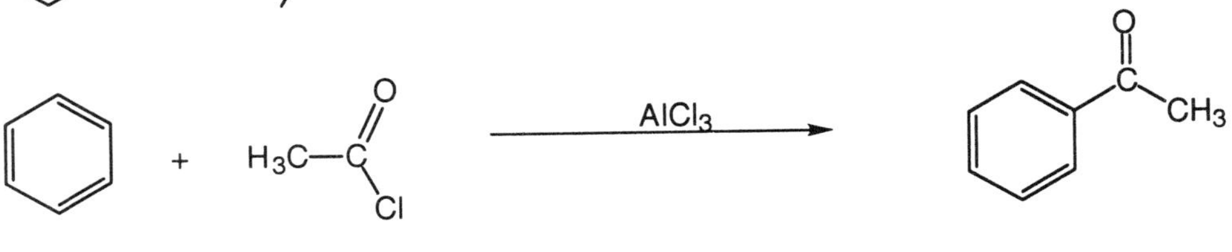

2. Consider the following reactions and explain the differences in reactivity and products.

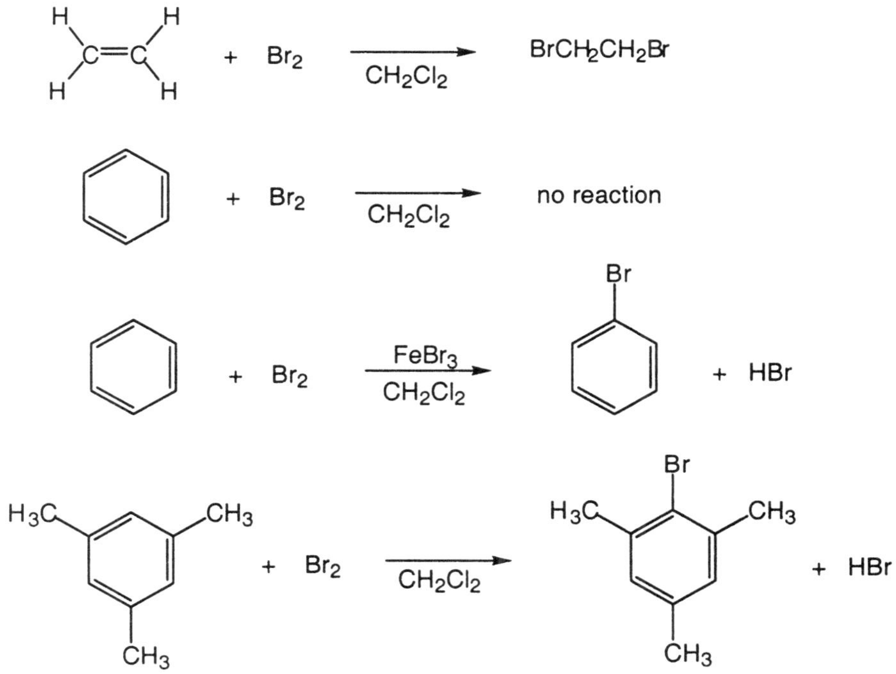

3. Explain clearly with words and structures the results of the following reactions. Which reaction is faster? Explain your choice.

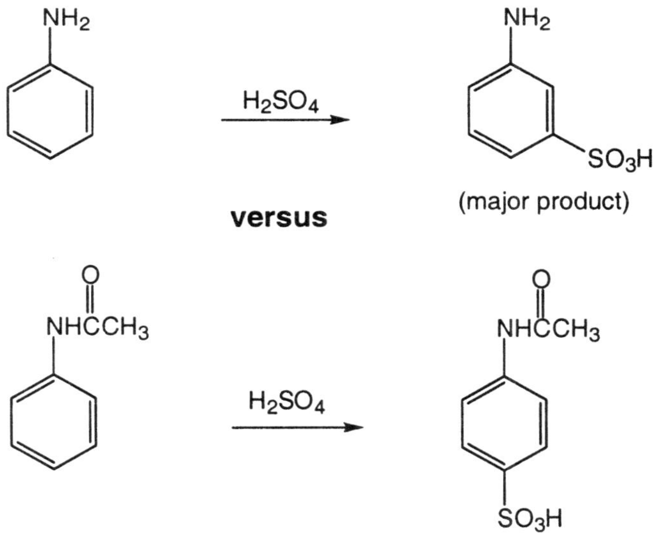

4. a. Predict which of the following reactions will occur at a faster rate.

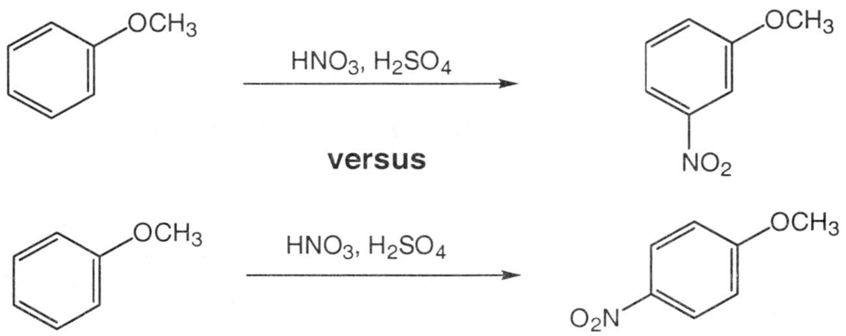

b. What does your answer in part (a) imply that the major product in the nitration of anisole ($PhOCH_3$) will be? Explain.

c. Show a detailed mechanism for the reactions in part (a) and give energy diagrams (reaction coordinate diagrams) for the two mechanisms. Take care to show all important resonance structures. Explain how your energy diagrams account for your predictions in parts (a) and (b).

5. Propose reagents and conditions for preparing the following compounds from benzene:

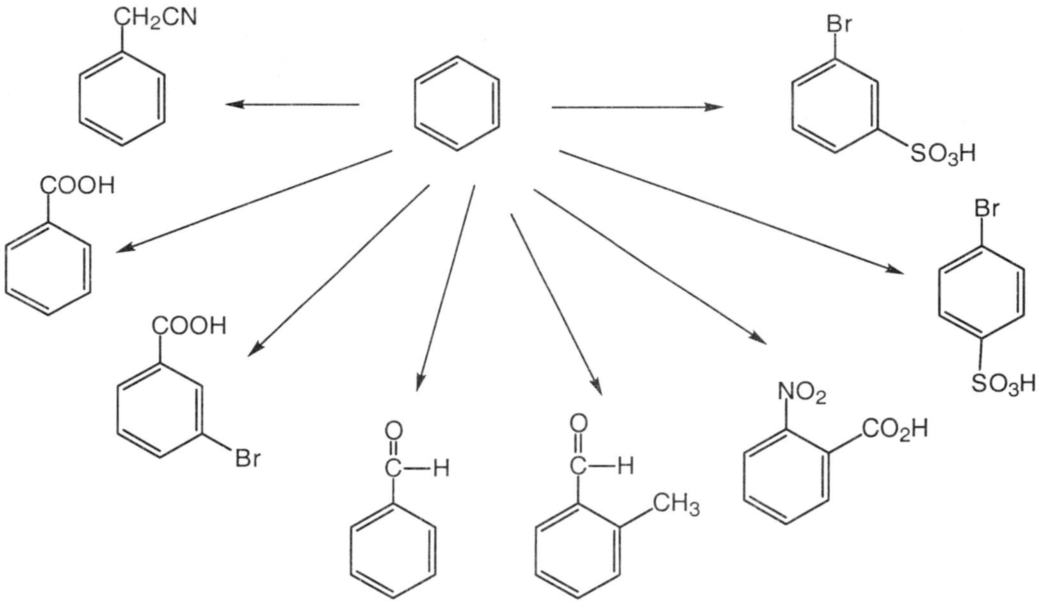

WORKSHOP GUIDE

Week 17

Aromatic Electrophilic Substitution

Problem 1 emphasizes that all the reactions listed have a common mechanism even though they appear to be different. This can be discussed using a round-robin format for the Workshop group. The Workshop leader can make the table on the blackboard while each student works through one reaction. The principle is that it is easier to learn one comprehensive mechanism than to learn seven isolated examples. If this is not apparent to the students, the leader must be certain to bring out the obvious connection to the mechanism of electrophilic addition to alkenes. In part 1(e), both mechanisms are operating!

Problem 2 emphasizes the mechanistic similarities to and differences between electrophilic addition to alkenes and electrophilic aromatic substitution. There are differences in rates and products. There are similarities in the rate-enhancing effect of the electron-releasing alkyl groups due to their ability to stabilize carbocation intermediates. The latter connects with the previous discussions about the relative stabilities of primary, secondary, and tertiary carbocations.

Problems 3 and *4* focus attention on directive effects in aromatic substitution. *Problem 4* dissects the orientation rules into questions of relative rates.

One of the reasons to learn the rules and concepts of aromatic substitution is to be able to make complicated molecules from simple precursors. *Problem 5* emphasizes how much can then be accomplished. The preparation of disubstituted benzenes illustrates the need to carefully arrange the order of the steps in a multistep sequence.

WORKSHOP

Week 18

Aldehydes and Ketones: Synthesis and Nucleophilic Additions

1. a. Discuss the similarities and differences in the structures and reactivities of the double bonds of 2-methylpropene and 2-propanone.

   b. With part (a) in mind, account for the following observations. Identify the nucleophile and the electrophile for the first step of each reaction.

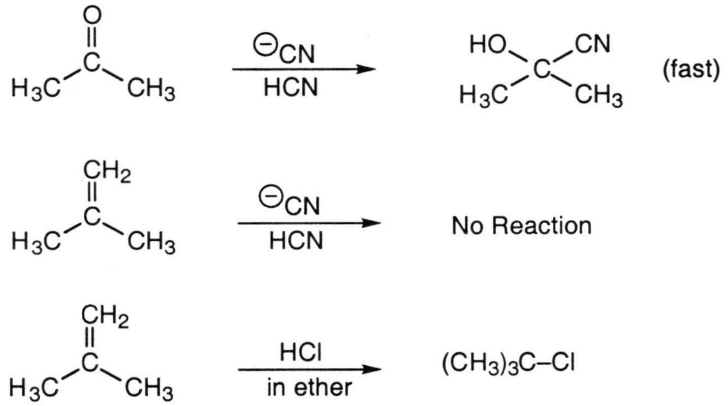

2. Write detailed electron-pushing mechanisms that explain the stated observations.

   a. The following reaction occurs rapidly at pH = 5 but fails at pH = 1.

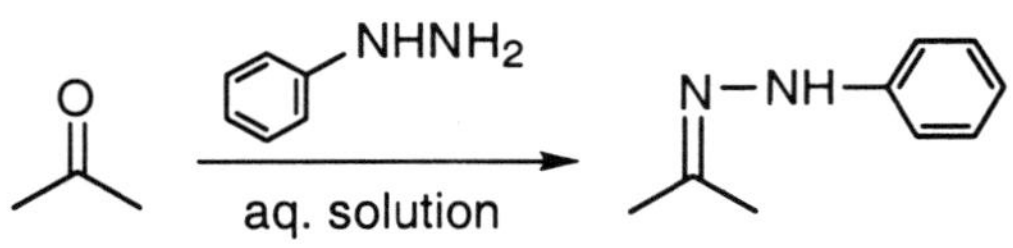

   b. The labeling of acetone with ^{18}O is catalyzed by both acid and base.

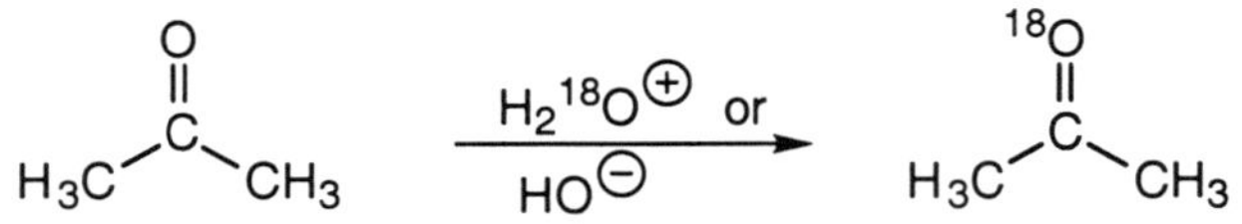

   c.

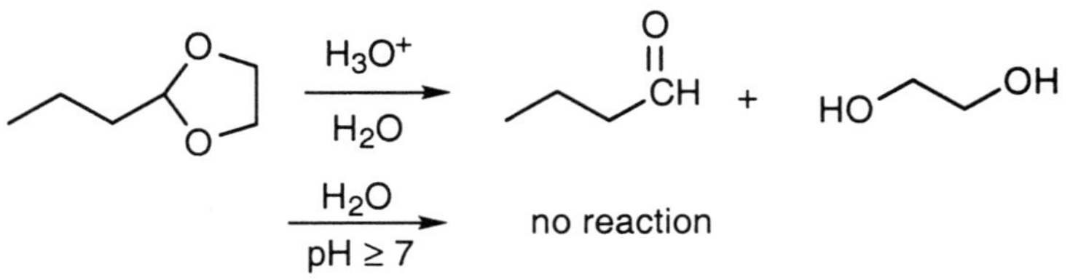

3. Propose structures for the lettered compounds.

a. Compound **X**, $C_7H_{14}O$, gives a silver mirror when treated with $Ag(NH_3)_2^{\oplus}/OH^{\ominus}$. Treatment of the compound with Zn(Hg)/HCl produces 2-methylhexane. The IR spectrum exhibits strong absorption at 1710 cm^{-1} and two moderate absorptions between 2700 and 2800 cm^{-1}. The ^1H NMR spectrum shows a one-hydrogen doublet at δ 9.70 (J = 2 Hz).

b. Compound **A**, $C_9H_{12}O$, was optically active, did not give a precipitate with 2,4-dinitrophenylhydrazine, showed a broad IR band at 3400 cm^{-1}, and was readily oxidized to **B**, $C_9H_{10}O$, with aqueous chromic acid (H_2CrO_4) at room temperature. When **A** was refluxed with chromic acid, benzoic acid was obtained. Compound **B** showed strong IR absorption at 1670 cm^{-1} but none at 3400 cm^{-1} and reacted with 2,4-dinitrophenylhydrazine reagent to give **C**. When **B** was reacted with EtMgBr followed by aqueous workup, compound **D** was obtained. **D** did not react with chromic acid reagent at room temperature and showed an IR band at 3400 cm^{-1}. **B** and **D** were optically inactive and could not be resolved.

4. Explain how to selectively carry out the following synthesis

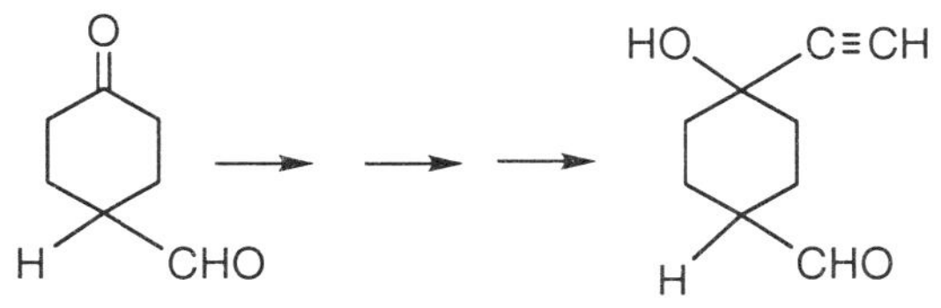

(mixture of cis and trans)

5. Show how to carry out the specified chemical conversions using any necessary organic and inorganic reagents. More than one step may be required.

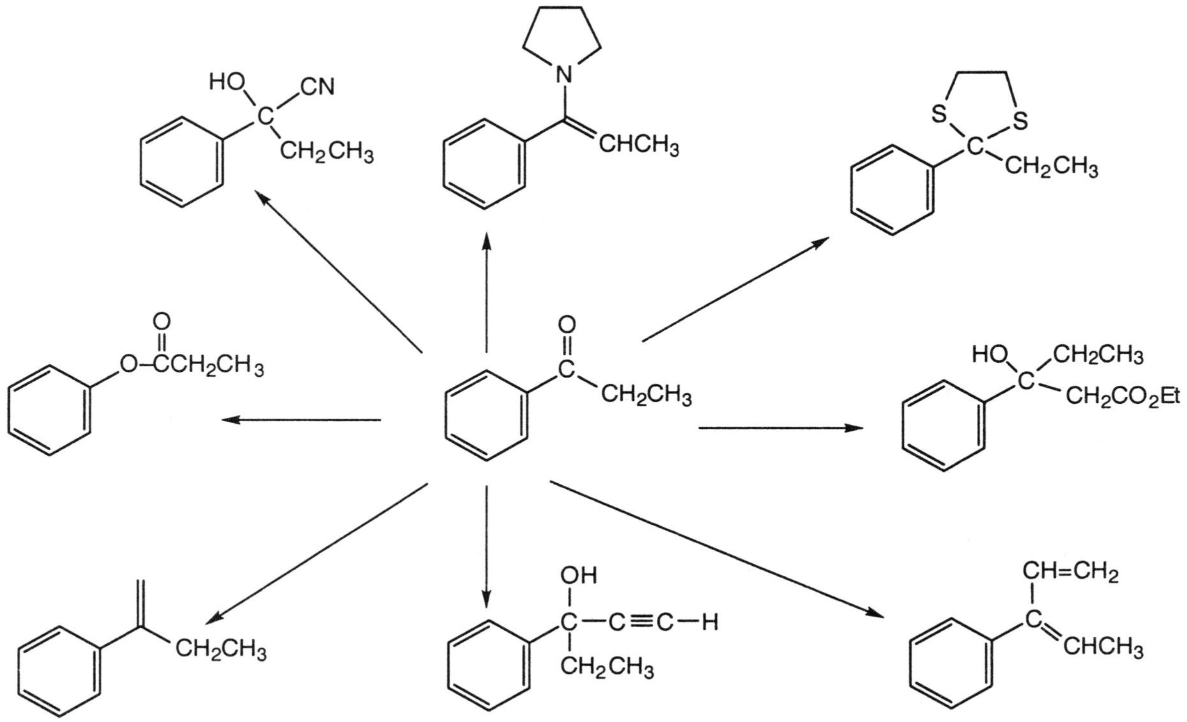

WORKSHOP GUIDE

Week 18

Aldehydes and Ketones: Synthesis and Nucleophilic Additions

Contemporary organic chemistry is marvelously coherent and, as a result, easy to learn. Our teaching, however, must bring out that coherence as an alternative to memorizing magical reactions. Simple considerations of structural differences (nonbonding electrons and bond polarization) between alkenes and carbonyl compounds lead directly to an understanding of the similarities and differences in reactions with electrophiles and nucleophiles. *Problem 1* focuses on the comparison of carbon-carbon and carbon-oxygen double bonds.

The "amphoteric" character of the carbon-oxygen double bond leads to important catalytic effects of acid and base on carbonyl additions. These are explored further in *Problem 2*. Before tackling the mechanism, the students should identify the electrophiles and nucleophiles in solution under different conditions.

Problem 3 returns to **Observation-Deduction** problems. We have, however, left out the prompt. If the students do not immediately break the problem into **Observation** and **Deduction** steps, the leader should remind them of the important of that logic.

Problem 4 shows what "faster-slower" understanding is good for, namely, for selective protection of functional groups in order to do selective synthetic transformations.

The carbonyl group is central to synthetic chemistry because of the ease and variety of reactions that form new bonds to the carbon of the C–O double bond. Most important of these is C–C bond formation. *Problem 5* emphasizes this bond-forming potential. It is useful to dissect the conversions in *Problem 5* to reveal the nature of the new bond to the carbonyl carbon.

WORKSHOP

Week 19

Aldehyde and Ketone Enol/Enolate Chemistry

1. Three compounds, **A**, **B**, and **C**, were all optically active, all had strong IR bands in the 1700 cm^{-1} region and all gave precipitates **D**, **E**, and **F**, respectively, with 2,4-dinitrophenylhydrazine reagent. When **A** was treated with $NaOD/D_2O$, three hydrogens were exchanged for deuterium (D), and the deuterated **A** remained optically active. When **A** was treated with I_2/aq. NaOH, a yellow precipitate (**G**) and **H** formed. Compounds **B** and **C** did not give a yellow precipitate with I_2/aq. NaOH. Compound **B** exchanged only one H for D when treated with $NaOD/D_2O$, and the monodeuterated **B** was then optically inactive. When compound **C** reacted with $NaOD/D_2O$, four H's were exchanged for D's, and the deuterated **C** remained optically active. Give structures for **A**–**H** consistent with the preceding information. (There may be more than one structure possible for **A**, **B**, or **C**.)

2. Give the products for the following reactions.

   a. $CH_3CH_2CH_2NO_2 + PhC(=O)H \xrightarrow[EtOH]{NaOEt}$

   b. $CH_3C(=O)CH_2CH_2C(=O)CH_3 \xrightarrow[H_2O]{^{\ominus}OH}$

   c. $CH_3CH_2C(=O)CH_2CH_3 + Br_2$ (1 mol) $\xrightarrow{HBr\ cat.}$

   d. $CH_3C(=O)Ph \xrightarrow{dil.\ aq.\ HCl\ cat.}$

3. Give a reasonable mechanism for each of the following reactions, clearly showing all important intermediates and resonance structures and using curved arrows to show movement of electron pairs.

   a. 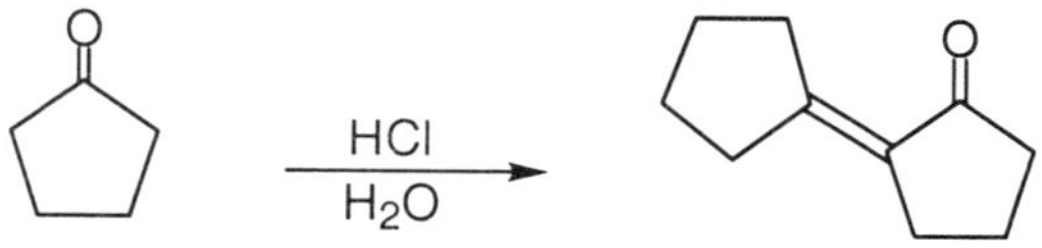

   b. 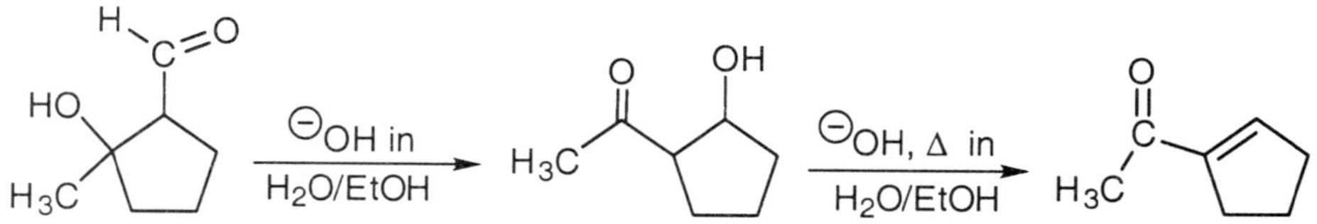

   c. (*R*)-2-methylcyclohexanone $\xrightarrow[in\ H_2O]{NaOH\ cat.}$ (*R,S*)-2-methylcyclohexanone

4. Show how to carry out the following chemical conversions using any necessary organic and inorganic reagents. More than one synthetic step may be required for a synthetic transformation.

a. 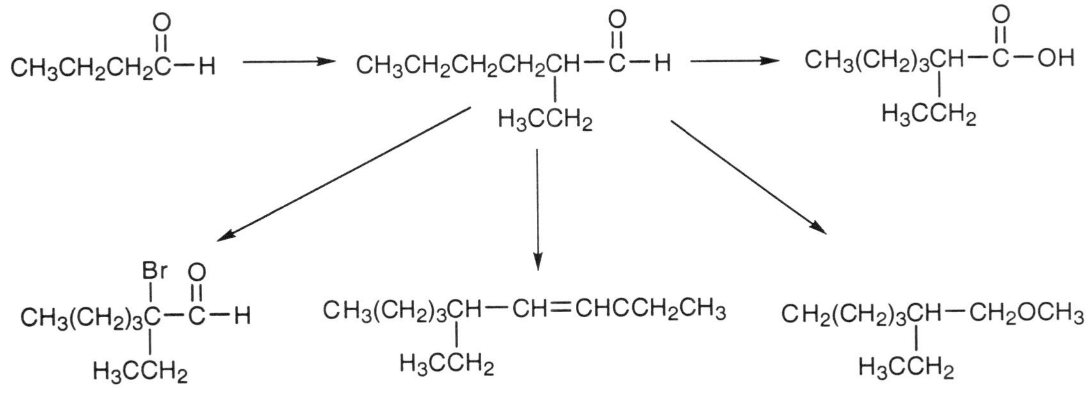

b. 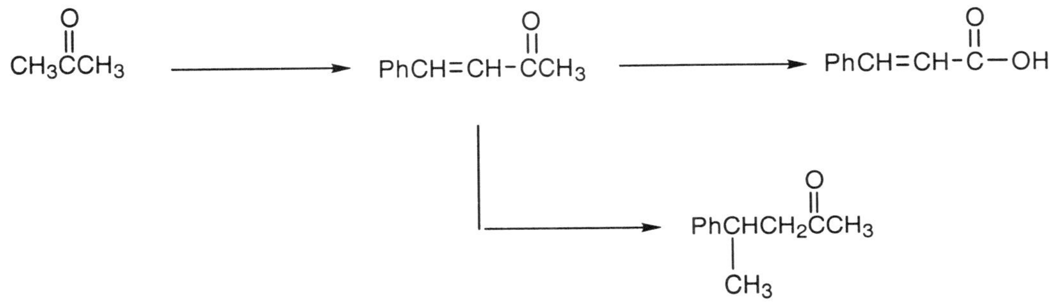

WORKSHOP GUIDE

Week 19

Aldehyde and Ketone Enol/Enolate Chemistry

Problems 1 and *2* are warm-up problems. These are intended to get the group started, and they also provide opportunities for the occasional less-than-well-prepared student to catch up. *Problem 1* is a good time for the leader to help the group recall the special characteristics of delocalized allyl-type systems.

Electron-pushing mechanism questions such as *Problem 3* are closely related to the thought patterns required in retrosynthetic analyses; they provide excellent practice in bond-breaking and bond-forming. Because the transformations really happen, the practice involves reactions that really happen. For example, there is a reverse-aldol lurking in *Problem 3(b).*

Problem 4 rings synthetic changes on aldol chemistry.

WORKSHOP

Week 20

Synthesis and Reactions of β-Dicarbonyl Compounds

1. a. Give a reasonable mechanism for the following reaction, clearly showing all important intermediates and resonance structures. Use curved arrows to show the movement of electron pairs.

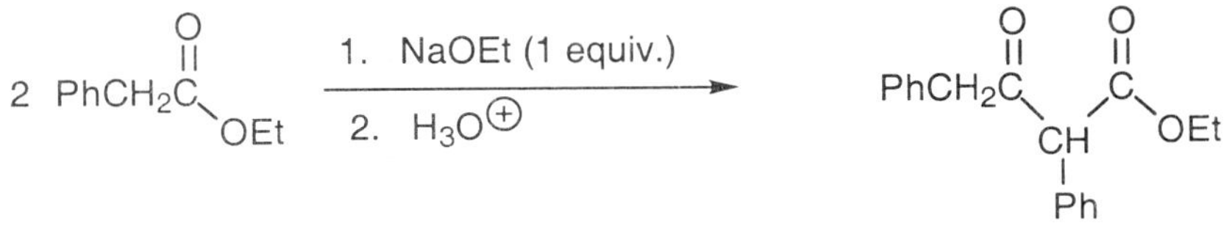

b. When compound **C** in ethanol is stirred with one molar equivalent of $NaOC_2H_5$ and one equivalent of $CH_3CH_2CH_2Br$, a new compound **D** is formed. When **D** is refluxed with 3N H_2SO_4, the ketone **E** is formed. Propose a structure for **D**, and give a stepwise electron-pushing mechanism leading from **C** to **D** to **E**.

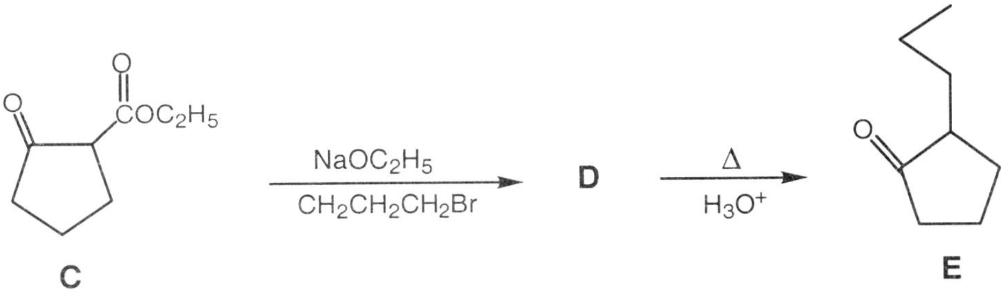

2. Provide structures for the major products of each of the following reactions. In each case, identify the electrophile and the nucleophile in the carbon-carbon bond-forming steps. Show the flow of electrons using the curved arrow formalism.

a.

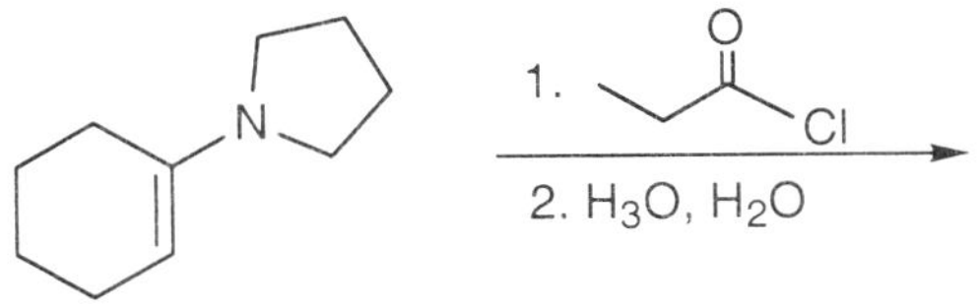

b.

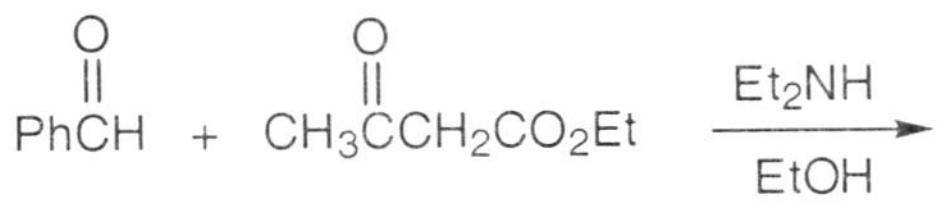

c. 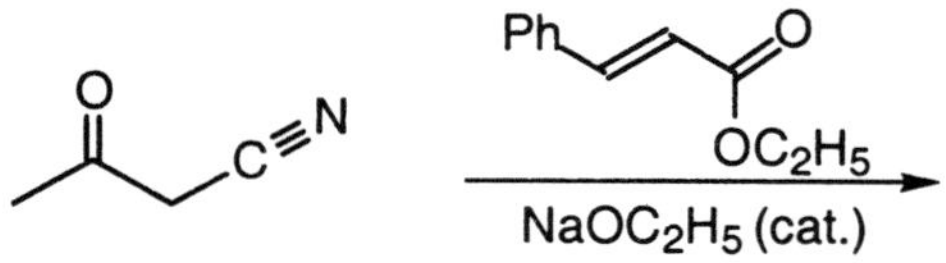

$(C_{15}H_{17}NO_3)$

d. $CH_3CH_2CO_2Et$ $\xrightarrow[\text{2 } CH_3CH_2I]{\text{1 } (Me_2CH)_2N^{\ominus}Li^{\oplus}}$

3. Disconnect the following compounds into components that would condense (assemble) to give the designated compounds. Use the electron-pushing formalism to show the condensation steps.

e.g., 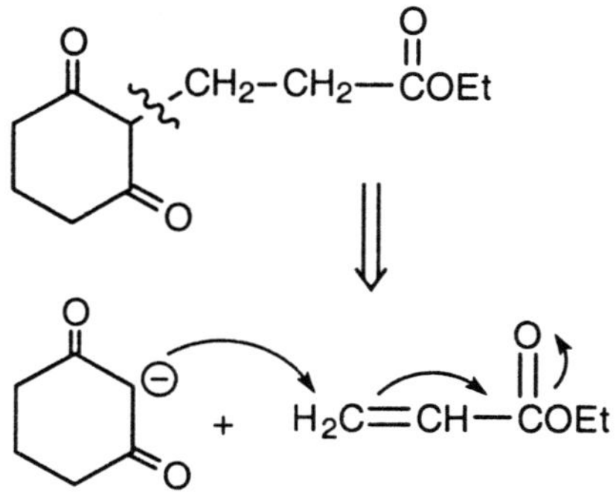

a. $C_6H_5C(=O)CH_2C(=O)OEt$ (two ways)

b. 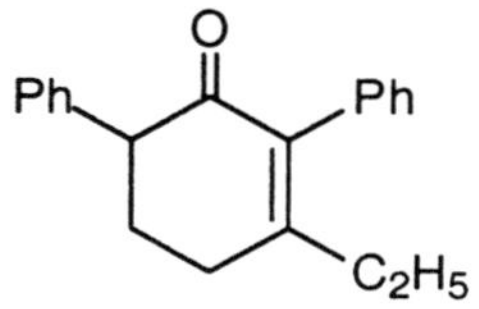

(*Hint*: MA says Robinson.)

c. $CH_3C(=O)CH(CH_3)CH_2CH{=}CH_2$

4. Having available ethyl acetoacetate, diethyl malonate, benzene, any compounds with three carbons or fewer, and any inorganic reagents, show how to synthesize the following compounds.

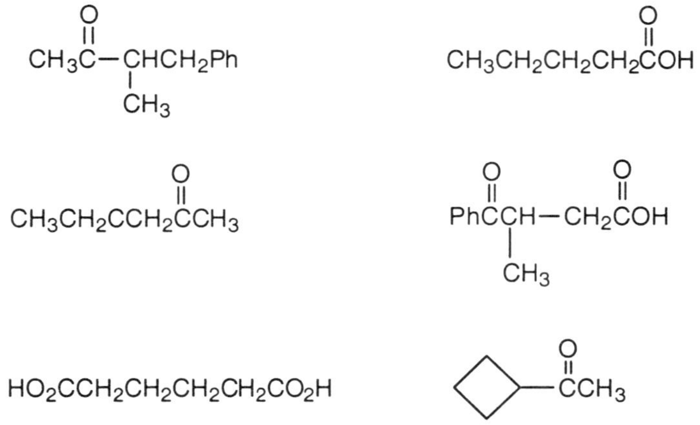

WORKSHOP GUIDE

Week 20

Synthesis and Reactions of β-Dicarbonyl Compounds

Although this Workshop emphasizes the synthesis and reactions of β-dicarbonyl compounds, some other reactions often covered with this material are also included (for example, Michael addition, enamine reactions, Robinson annulation). The major challenge in working with these important synthetic reactions is to be able to "see" the component pieces in the final structure. This Workshop leads the students through a graded series of steps from known reactions to less familiar reactions to retrosynthesis to synthesis. Identifying the C–C bond-forming reaction of nucleophile with electrophile should cause patterns of sense to emerge. A method of analysis of synthetic problems should also emerge.

Good leaders are "can-do" types who show others that they "can do" also. This Workshop makes special demands on the leader's abilities to guide, support, and encourage.

WORKSHOP

Week 21

Carboxylic Acids

1. Specify the member of each of the following pairs that is more acidic. Explain your choice in words and with the help of structures.

a. CH_3COOH versus FCH_2COOH

b. CF_3COOH versus CCl_3COOH

c. $CH_2{=}CHCH_2COOH$ versus $HC{\equiv}CCH_2COOH$

d. $N{\equiv}C{-}CH_2COOH$ versus $HC{\equiv}CCH_2COOH$

e. 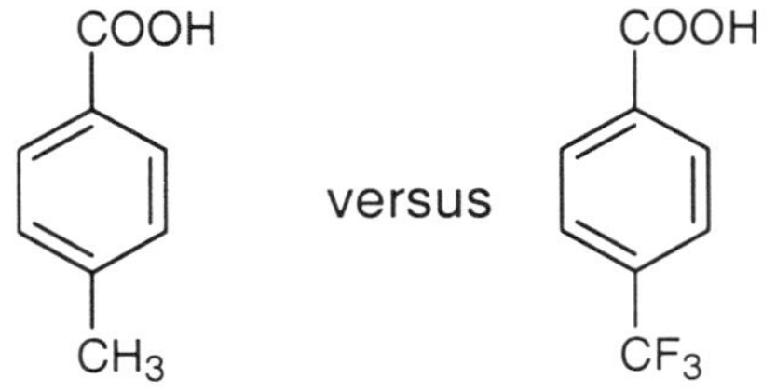

f. 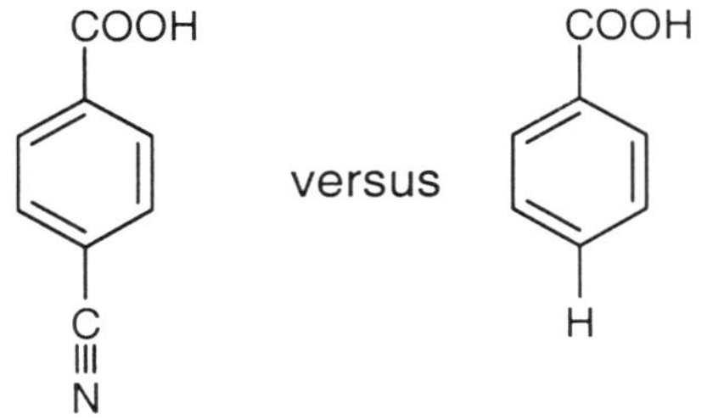

2. Predict the relative rates of electrophilic aromatic substitution for substituted arenes, using the same substituents as in Problem 1(e) and 1(f); that is, which member of the following pairs reacts faster?

a. 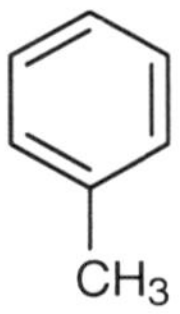 versus 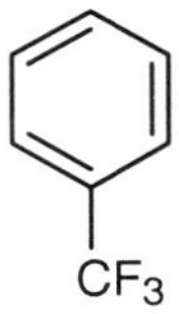

b. 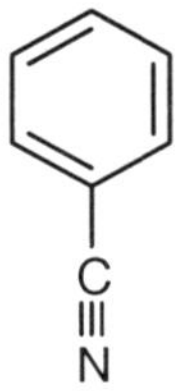 versus 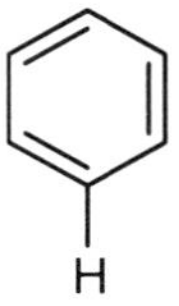

How do the effects of substituent on the acidities of benzoic acids compare with the rates of electrophilic substitution? Formulate a predictive "rule of thumb." Explain the structural basis for your rule.

3. Predict the major species at equilibrium. Justify your choices.

a. $C_6H_5CO_2H + H_2O \rightleftharpoons C_6H_5CO_2^{\ominus} + H_3O^{\oplus}$

b. $CH_3CO_2^{\ominus} + EtO_2CCH_2CO_2Et \rightleftharpoons CH_3CO_2H + EtO_2C\overset{\ominus}{C}HCO_2Et$

c. $CH_3CH{=}C(OH)_2 \rightleftharpoons CH_3CH_2C(=O)OH$

4. a. Using words and equations, explain the observations in each of the following reactions:

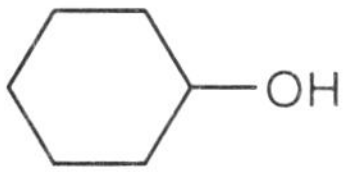

+ NaOH (aq) ⟶ insoluble

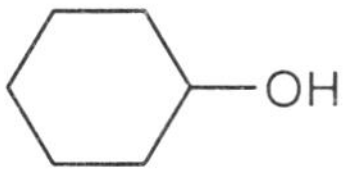

+ $NaHCO_3$ (aq) ⟶ insoluble

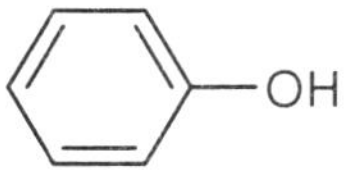

+ NaOH(aq) ⟶ soluble

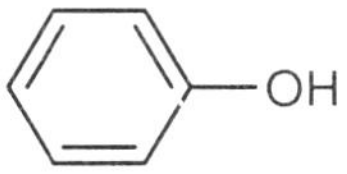

+ $NaHCO_3$ (aq) ⟶ insoluble

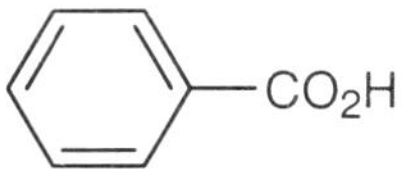

+ NaOH (aq) ⟶ soluble

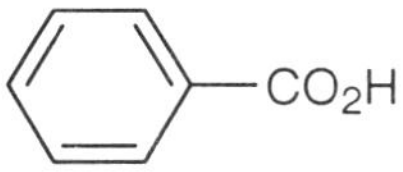

+ $NaHCO_3$ (aq) ⟶ soluble, formation of gas bubbles

b. Using the results of the reactions in part (a), provide a flow diagram showing how products of hydrolysis of each of the following esters with aqueous NaOH can be separated by acid-base extraction techniques.

i. 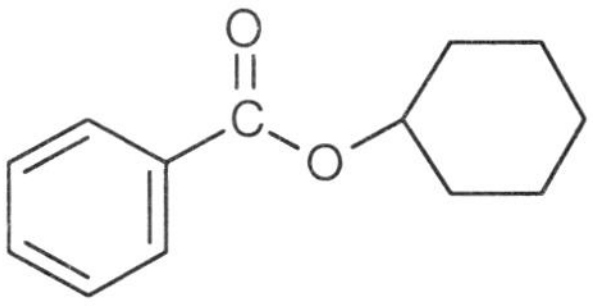

ii. 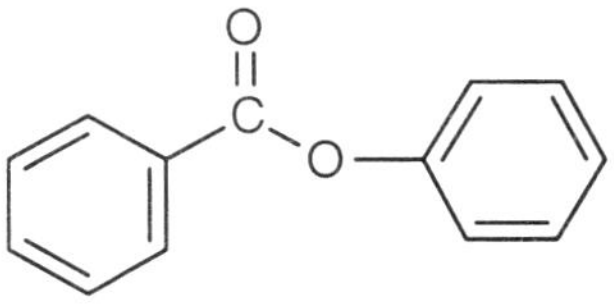

5. Show how to carry out the following synthetic conversions. Specify necessary reagents and conditions.

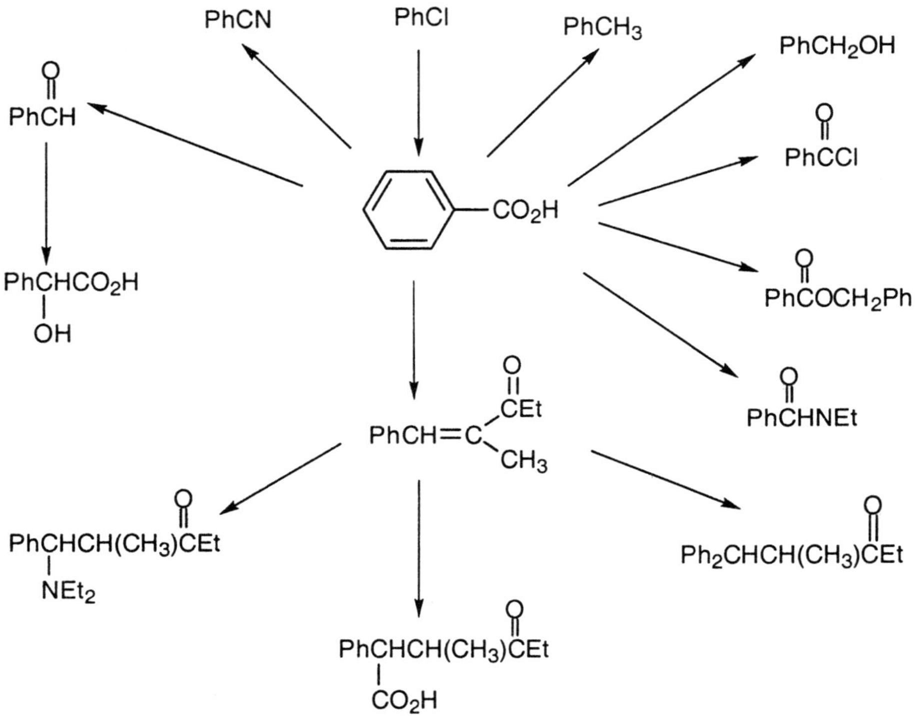

WORKSHOP GUIDE

Week 21

Carboxylic Acids

Many of the fundamental ideas about the nature and effects of substituents on rates and equilibria are grounded in the effects of substituents on the acidities of carboxylic acids. The first two problems in this Workshop draw attention to one of the classic examples of the connection between equilibrium and rate. The leader might ask the students to predict effects of substituents on other rates, for example, rate of electrophilic addition to substituted alkenes.

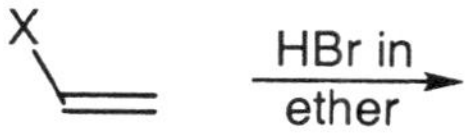

Problems 3(a), (b) and *(4)* are related to the position of equilibrium in various acid-base reactions. In spite of considerable experience with acid-base problems, there always seems to be a gap between theory and practice. These problems move from theory to practice: just what is the dominant species at equilibrium? How does the equilibrium affect a macroscopic observable such as solubility? How can our understanding of these equilibria be put to practical use to solve real problems?

Problem 5 explores some of the rich preparative chemistry of the carboxylic acid functionality. As always, it is best to start these transformations by identifying what needs to be accomplished. That identification will, in turn, lead to the right choice of reagents and reactants. For example, the leader might start by asking the students to find all the redox transformations in the scheme, which means that, sooner or later, redox reagents will be required. Identification of the C–C bond forming transformations will lead to a different set of reagents. It is always useful to distinguish functional group interchanges and C–C bond-forming reactions. Of course, it is also always useful to think retrosynthetically.

WORKSHOP

Week 22

Carboxylic Acid Derivatives: Nucleophilic Acyl Substitutions

1. Each of the following reactions involves nucleophilic substitution at an acyl carbon (carbonyl of a carboxylic acid or derivative) by way of a tetrahedral addition intermediate. Construct a table showing the electrophile and the nucleophile in the key bond-forming step and the corresponding tetrahedral addition intermediate for each reaction. Write a general mechanistic scheme (use curved arrows to show movement of electron pairs) for these reactions, keeping in mind that acid- and base-catalyzed processes will differ in timing of proton transfers.

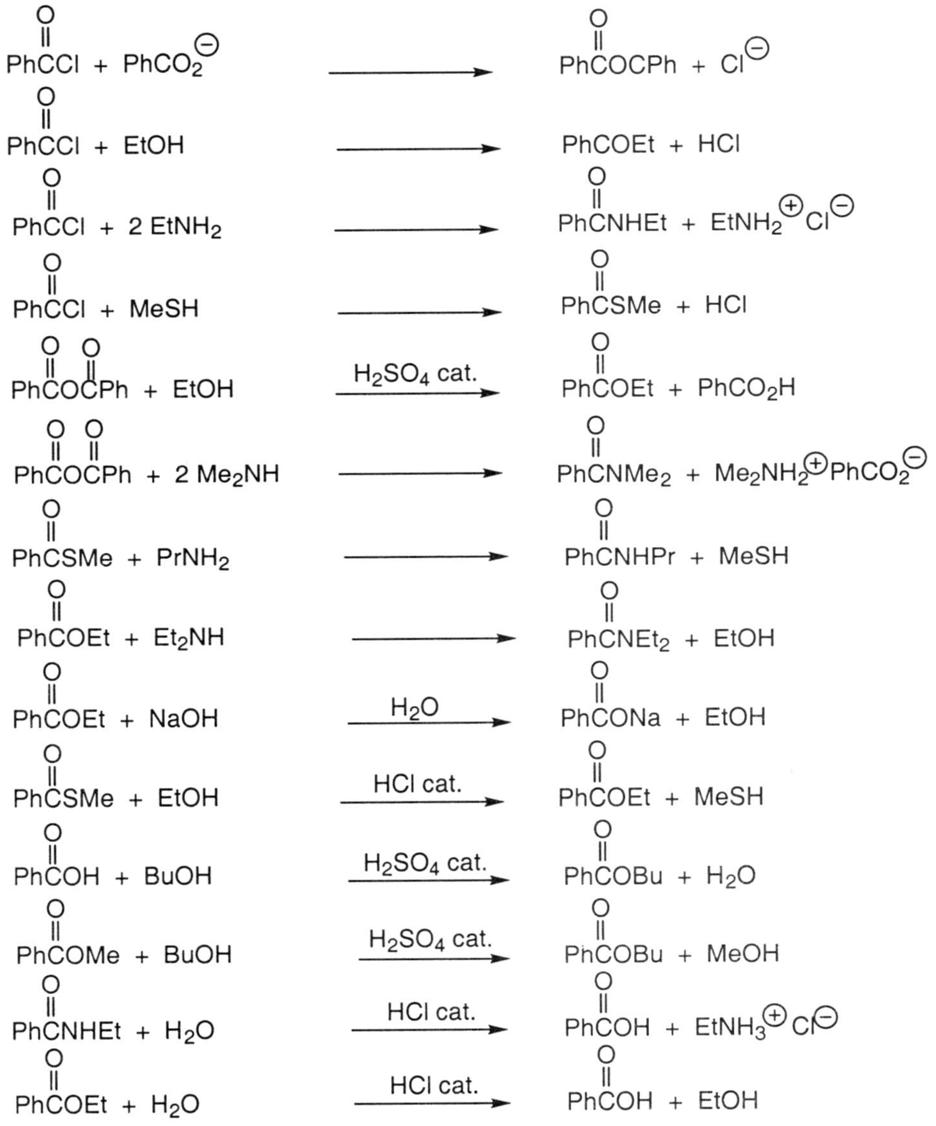

2. a. In the following case, predict which of the two competing reactions will be faster, and explain your choice.

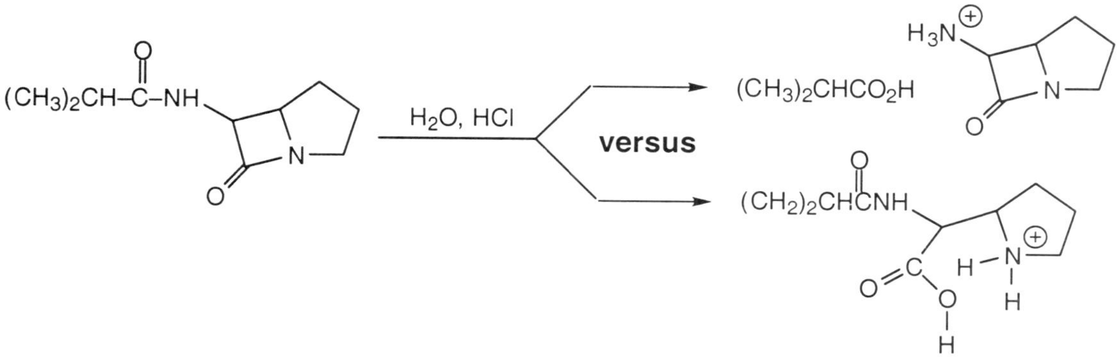

b. Propose an explanation for the observation that the bromination of the acyl bromide is preparatively useful, whereas the bromination of the acid is not.

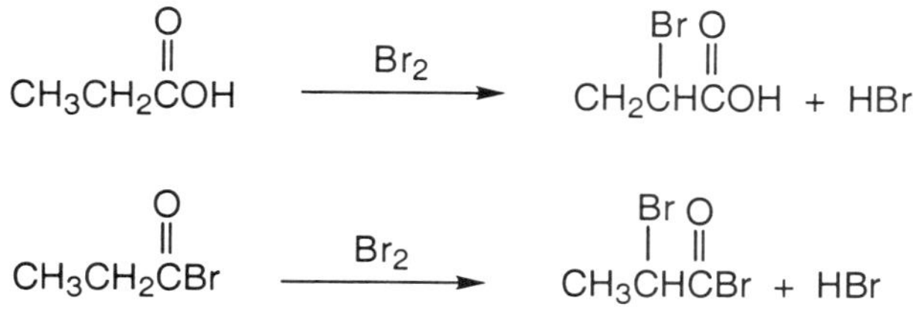

3. The ester shown is labeled with oxygen-18 as indicated (*O=^{18}O). Give a mechanism consistent with the labeling results shown when the ester is hydrolyzed in unlabeled water at pH = 2.

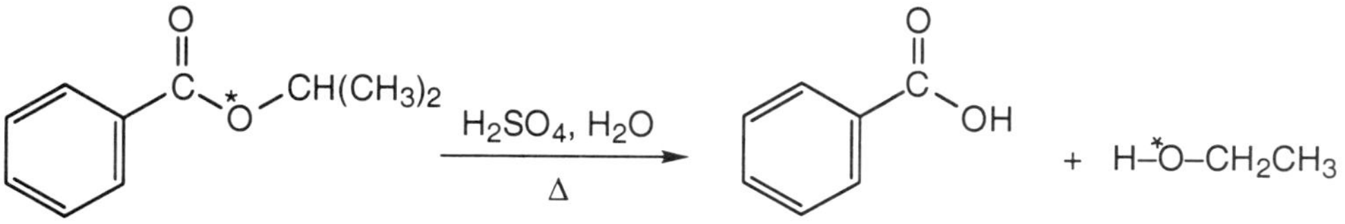

4. Compound **A**, $C_{12}H_{16}O_2$, exhibited a strong absorption in the IR spectrum at 1715 cm^{-1}. Treatment of **A** with $NaOH/H_2O$ gave a neutral compound **B** and a sodium salt that after acidification gave compound **C**. **B** exhibited strong broad absorption in the IR spectrum at 3333 cm^{-1} and could be oxidized with aqueous chromic acid ($Na_2Cr_2O_7/H_2SO_4/H_2O$) to 2-methylbutanoic acid. **C** exhibited absorption in the IR spectrum at 1681 cm^{-1} and a very broad band over the range 2500-3500 cm^{-1}. The ^1H NMR spectrum of **C** exhibited absorption at δ 7.1-8.5 (m, 5H) and 12.70 (s 1H). Provide structures for **A**, **B**, and **C**.

5. Derivatives of carboxylic acids are important in many technologies. Consider the following applications.

a. Benzyloxychloroformate ($BO_2C–Cl$), $PhCH_2O\overset{O}{\overset{\|}{C}}Cl$, is an important reagent used in the synthesis of peptides and proteins as a protective group for the amino groups of aminoacids.

i. Show how BO_2C-Cl could be synthesized from phosgene.

ii. Give a structure for the product of reaction of BO_2C-Cl with the amino acid glycine, $H_2NCH_2CO_2H$.

b. 4-Aminobutanoic acid, also known as GABA (γ-aminobutyric acid), is an agent in the transmission of nerve impulses. The compound has been synthesized from 4-chlorobutanenitrile ($ClCH_2CH_2CH_2CN$) using the Gabriel (phthalimide) synthesis. Provide chemical equations showing how the synthesis could be accomplished.

c. Dacron is a polyester formed by transesterification of dimethyl terephthalate with ethylene glycol to give a long-chain polymer molecule. Provide a partial structure of Dacron that shows two of the repeating units of the polymer.

dimethyl terephthalate

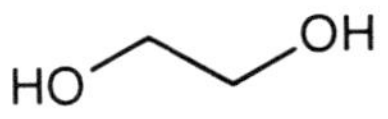

ethylene glycol

WORKSHOP

Week 22

Carboxylic Acid Derivatives: Nucleophilic Acyl Substitutions

By now, the point of *Problem 1* should be obvious: it is easier to learn reactions that are connected parts of a general mechanistic scheme. Once the group has developed the general scheme, a round-robin approach to the individual examples in the diverse set of reactions is appropriate.

Faster-slower choices are central to an understanding of the structure-reactivity interplay. When the mechanism of the reaction is known, it is the key to predicting and understanding relative rates. Conversely, a formulation of the mechanism is often based on observations of rate as a function of structure. *Problem 2* in this Workshop requires a knowledge of the mechanism of the reaction in question. There is a faster-slower problem built into *Problem 1* as well. The leader could ask the students to order the reactivities for different acyl derivatives with the same nucleophile; for example, acid chloride, anhydride, ester, thioester, and amide with H_3N.

Problem 3, on the other hand, reverses the logic and asks the student to choose a mechanism on the basis of relative rate data. To make this thought process clearer, the leader should ask the group to propose two different mechanisms corresponding to the cleavage of two different C–O bonds. These two mechanisms (i.e., two hypotheses) are distinguished by the labeling test; the labeling reveals that one path (mechanism) is faster and the other slower.

Problems 4 and *5* are reminders that once one knows something well, one can do something useful. All structure-proof problems, such as *Problem 4*, are about extracting reliable conclusions from experimental observations. *Problem 5* shows how an understanding can be put to work to make useful and interesting reagents, compounds, and materials.

WORKSHOP

Week 23

Amines

1. a. The following compounds are stored in two unlabeled bottles. Both compounds are water insoluble. Using a simple solubility test, how would you determine which bottle contains which compound? Give the chemical reaction for the positive test. What other chemical reaction might be used to differentiate these two compounds? Which spectroscopic method could be used to distinguish the two compounds? Explain exactly what you would do and exactly what you would expect to observe for the two different compounds in each test. Give equations for the chemical reactions that correspond to the "positive" tests.

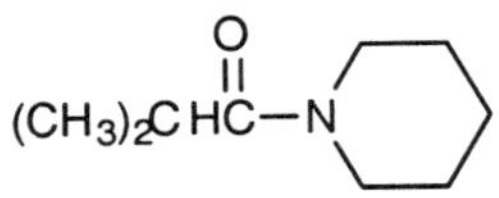

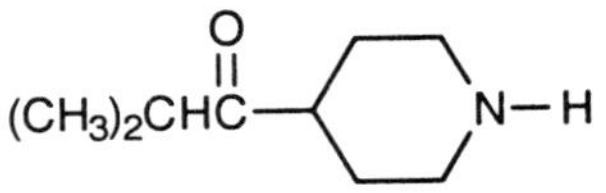

b. Provide the structure of Novocain, which is the salt formed from Procaine (structure shown) and one equivalent of hydrogen chloride. Explain the basis of your choice.

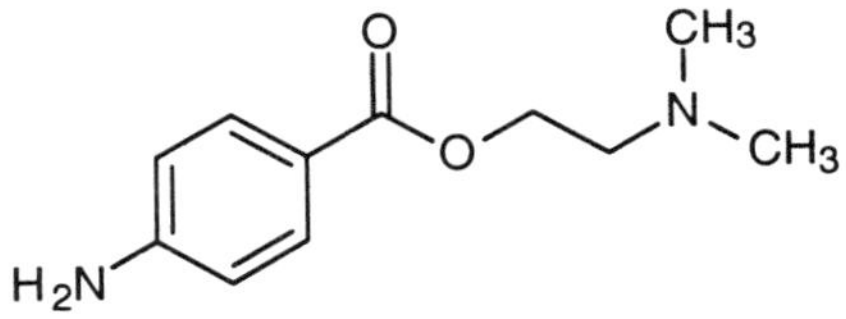

2. a. Explain, in words and structures, the conceptual basis of the separation of a racemic mixture into its enantiomeric components.

b. Draw flow charts to show two different ways that you could use an optically active amine, (+)–RNH_2, to resolve a racemic mixture of an acid, (±)–RCOOH. Explain your choice of reagents and reaction conditions at each step.

3. Give structures for compounds **S** through **W** and specify missing reagents in the following scheme.

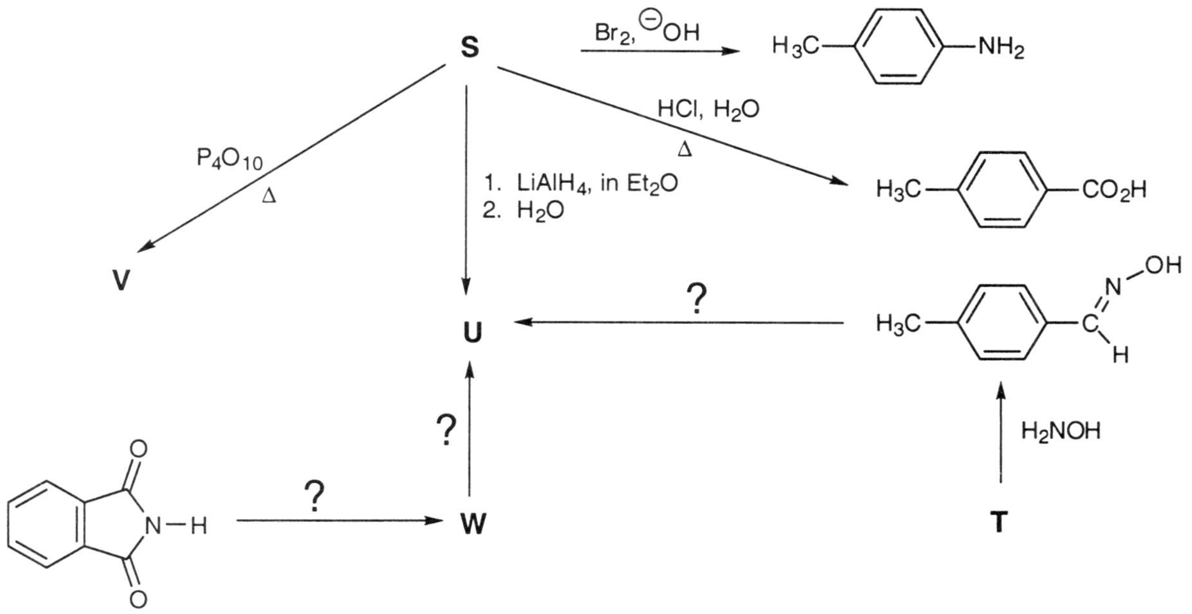

4. Provide reagents necessary to bring about each of the conversions indicated by an arrow.

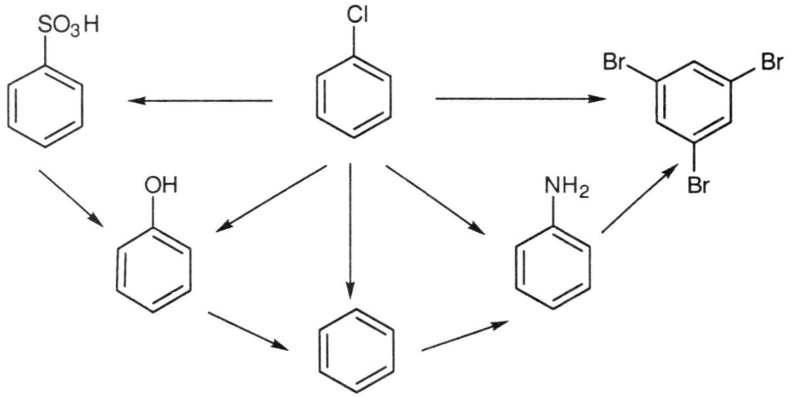

5. The following amide gives different products depending on whether it is reacted with bromine and aqueous NaOH or bromine and NaOMe/MeOH. In each case, optical activity in the starting materials is preserved in the products.

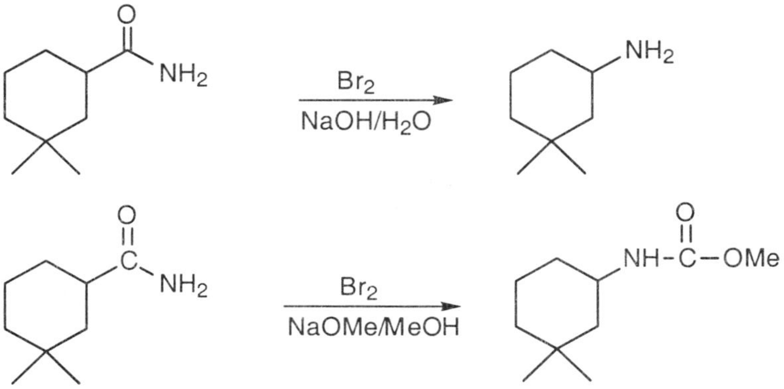

a. Identify which bonds are broken and which are made in the course of these rearrangements.

b. Identify the stereogenic centers in the reactants and products.

c. Give electron-pushing mechanisms for the two reactions. Identify the electrophile and nucleophile in each step in the mechanism. What is the role of the Br_2? What is the role of the bases ($^{\ominus}$OH or $^{\ominus}$OMe)?

d. Explain the origin of the difference in products in the two reactions.

WORKSHOP GUIDE

Week 23

Amines

Problem 1 in this Workshop deals with basicity of nitrogen-containing compounds. It will be helpful if the Workshop leader asks the students to identify the functional groups present in the two compounds and talk about the different properties of these groups before identifying the means to distinguish them.

The leader will want to start *Problem 2* by reviewing the ideas behind the words and the notations: optically active, resolve, racemic mixture, (+)-, (±)-. The problem will take the students back to basic ideas about enantiomers and diastereomers and provides an opportunity to recall that the properties of diastereomers (for example, solubilities, rates of reaction) are different. The different resolution schemes make use of different chemical reactions of amines and carboxylic acids (or their derivatives).

Problem 3 is a synthesis problem disguised as a puzzle problem. As such, it helps to think in both directions, from reactants to products, and vice-versa. *Problem 4* is the more obvious synthesis problem, and the principle obviously applies to *Problem 4* as well. *Problems 3* and *4* segregate the chemistry of amines into aliphatic and aromatic sets. It is helpful to get students to talk about the differences in structure, properties, and reactivity [compare with *Problem 1(b).*

Our Workshop leaders are insistent about two key tactics. The first is that it is best to limit the number of Workshop ideas in order to allow time to explore them thoroughly. The second is that the way to deal with a difficult reaction or concept is to structure the problem to lead the students step-by-step to understanding. Taking this advice to heart means that this Workshop on amines is too long. It is deliberately so, however, in order to include a good example of leading the students to understanding. *Problem 5* could have been written as: *Explain the product difference.* The buildup to that final question makes the problem manageable and helps build confidence and understanding.

WORKSHOP

Week 24

Nucleophilic Aromatic Substitution

1. a. The pK_a of CH_3CH_3 is 50-60 and the pK_a for CH_3NO_2 is 10. Explain, in words and structures, the origin of this difference.

   b. How does this pK_a difference help to explain the difference in the reactivities in the following reactions? Make your answer explicit by giving an arrow-pushing mechanism for the substitution reaction.

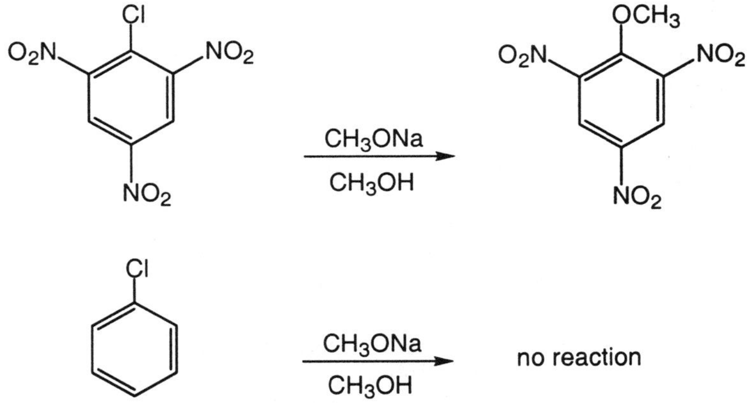

2. a. List the following compounds in order of increasing reactivity with NaOEt/EtOH. Explain how you made your choices.

   b. For the most reactive compound of the series, give a reaction mechanism that specifies the structure of any intermediates and the reaction product.

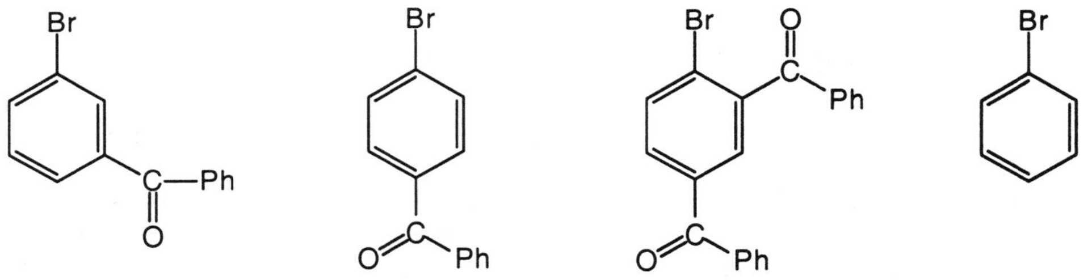

3. Examine the following pairs of reactions. Explain, with the help of words and structures, the observed rate differences.

a.

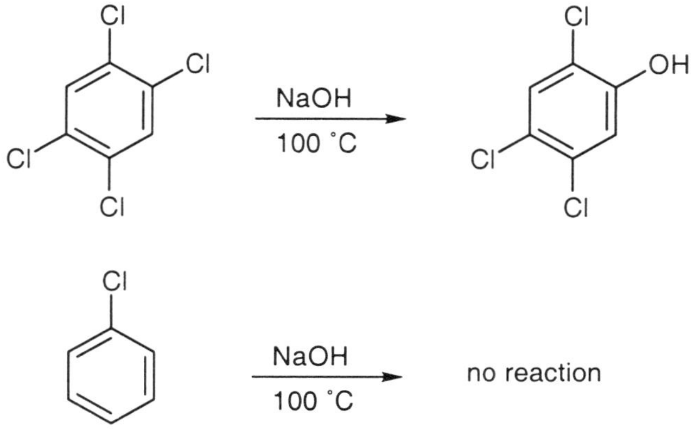

b.

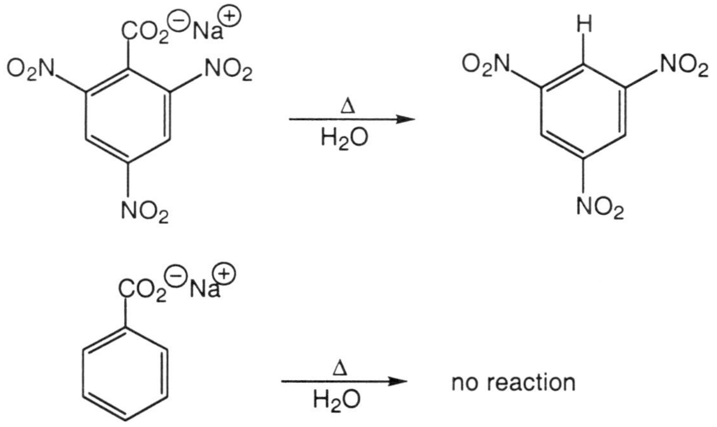

4. Provide the structure of the dimethylbromobenzene isomers that satisfy the following descriptions:

a. Does not react with KNH_2 in liquid ammonia.

b. Gives only one product when it reacts with KNH_2 in liquid ammonia.

c. There are four other dimethylbromobenzenes. Give their structures. Two of these (**1** or **2**) react with KNH_2/NH_3 to give the identical mixture of two dimethylanilines. Two other dimethylbromobenzenes (**3** or **4**) also react separately to give the identical mixture of dimethylanilines, but these dimethylanilines are different from those derived from compounds **1** and **2**. Explain clearly why the identical mixture of dimethylanilines is formed from two different starting materials (for example, **1** or **2**). Also explain clearly why there is no crossover in the dimethylanilines formed from **1** or **2** and **3** or **4**, respectively.

5. Suggest mechanisms for the following reactions.

a. 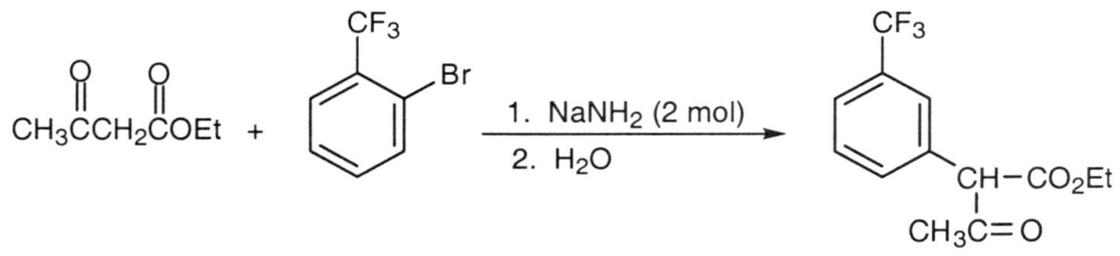

b. 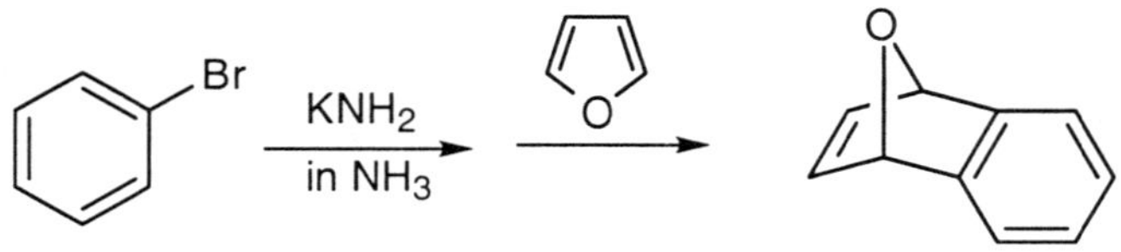

WORKSHOP GUIDE

Week 24

Nucleophilic Aromatic Substitution

It is very important to connect the reactions in this Workshop to previous ideas and reactions. The first problem sets the stage by reminding students of the resonance stabilization of the conjugate base of nitromethane. The second problem is parallel and should connect back to previous study on enolate ions. This is a good time for the leader to ask about the pK_a of acetone (or acetophenone) compared with that of ethane.

Charge can be stabilized by resonance *or* by inductive effects. *Problem 3* explores inductive effects. The leader might ask about the effects of NO_2 and Cl groups on the pK_a's of substituted acetic acids or *p*-substituted benzoic acids. A connection back to Workshop 21 on carboxylic acids would be helpful.

Some of our confidence about the benzyne intermediate comes from the "symmetrizing" character of the elimination reaction. *Problem 4* explores this idea. The problem will work backward as well. The different product results reveal that the adjacent carbons must somehow be involved in the mechanism of the substitution reactions.

A classic question in the alkyne chapter asks about the stability of small-ring alkynes such as cycloheptyne and cyclooctyne. At that stage, the point is to get students to work from their knowledge that the geometry of alkynes is linear to the energetic consequences of bending an alkyne. More sophisticated analysis would consider what happens to the bonding as the alkyne is bent. The structure and reactivity of benzyne provides an opportunity to revisit these ideas, helping the student to see that the chemistry of benzyne makes sense, even if it remains surprising.

WORKSHOP

Week 25

Phenols

1. Compounds **A**, **B**, and **C** all have a molecular formula of $C_{10}H_{12}O_2$, and their IR and NMR spectra show that they are all *p*-disubstituted benzenes. All of these compounds are water insoluble. Compound **A** is insoluble in both NaOH and $NaHCO_3$. Compound **B** is soluble in both NaOH and $NaHCO_3$, and compound **C** is soluble in NaOH but insoluble in $NaHCO_3$. Give structures for **A**, **B**, and **C**, consistent with this information.

2. a. Show how each of the following compounds can efficiently be converted into phenol.

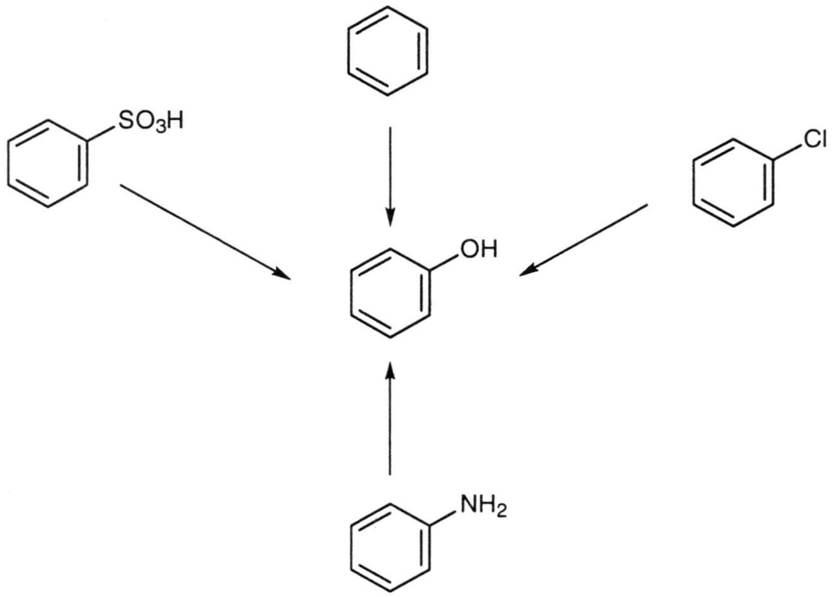

b. Suppose that you reasoned by analogy and tried to make *p*-methylphenol from *p*-chlorotoluene, *p*-aminotoluene, and *p*-toluenesulfonic acid using the methods you employed in part (a) for the monosubstituted benzenes. Predict the results of the three experiments.

3. Give a reasonable mechanism for each the following reactions. Use curved arrows to show movement of electron pairs and "fish hooks" to show the movement of single electrons.

a. 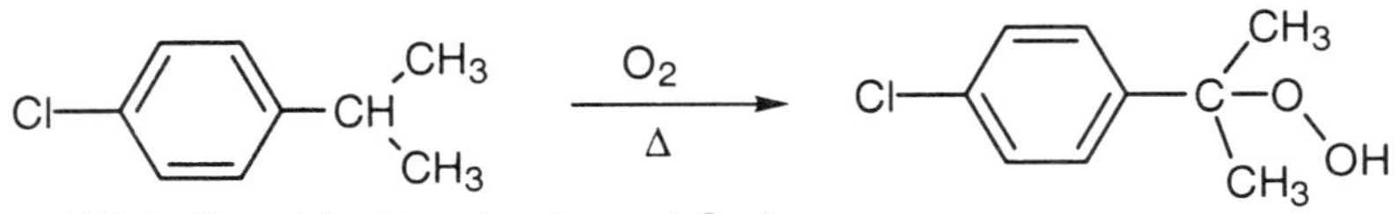

(*Hint*: Consider the structure of O_2.)

b. 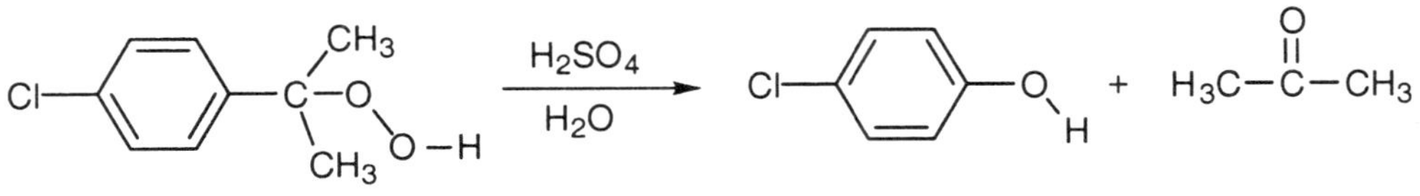

c. 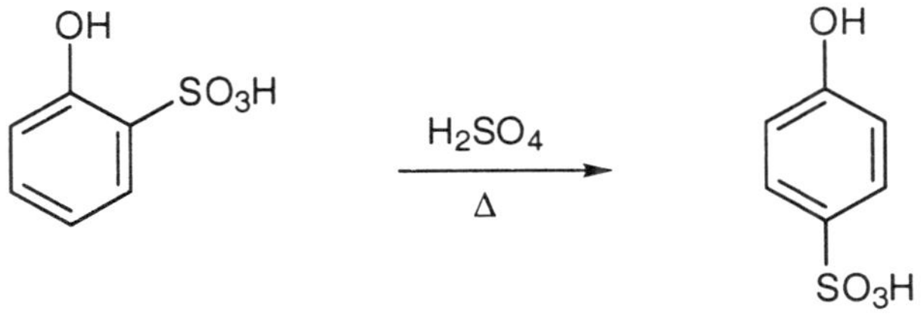

d. 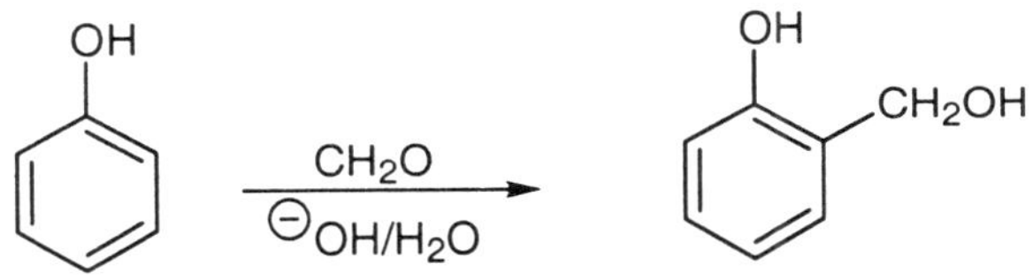

e. 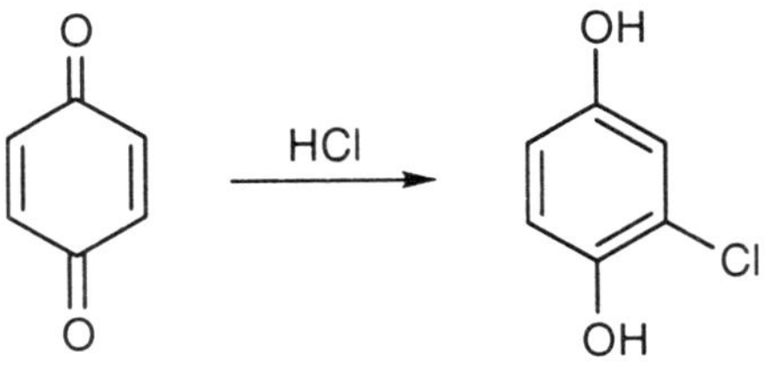

4. Show how to carry out the following synthetic conversions efficiently and in good yield using any necessary organic and inorganic reagents.

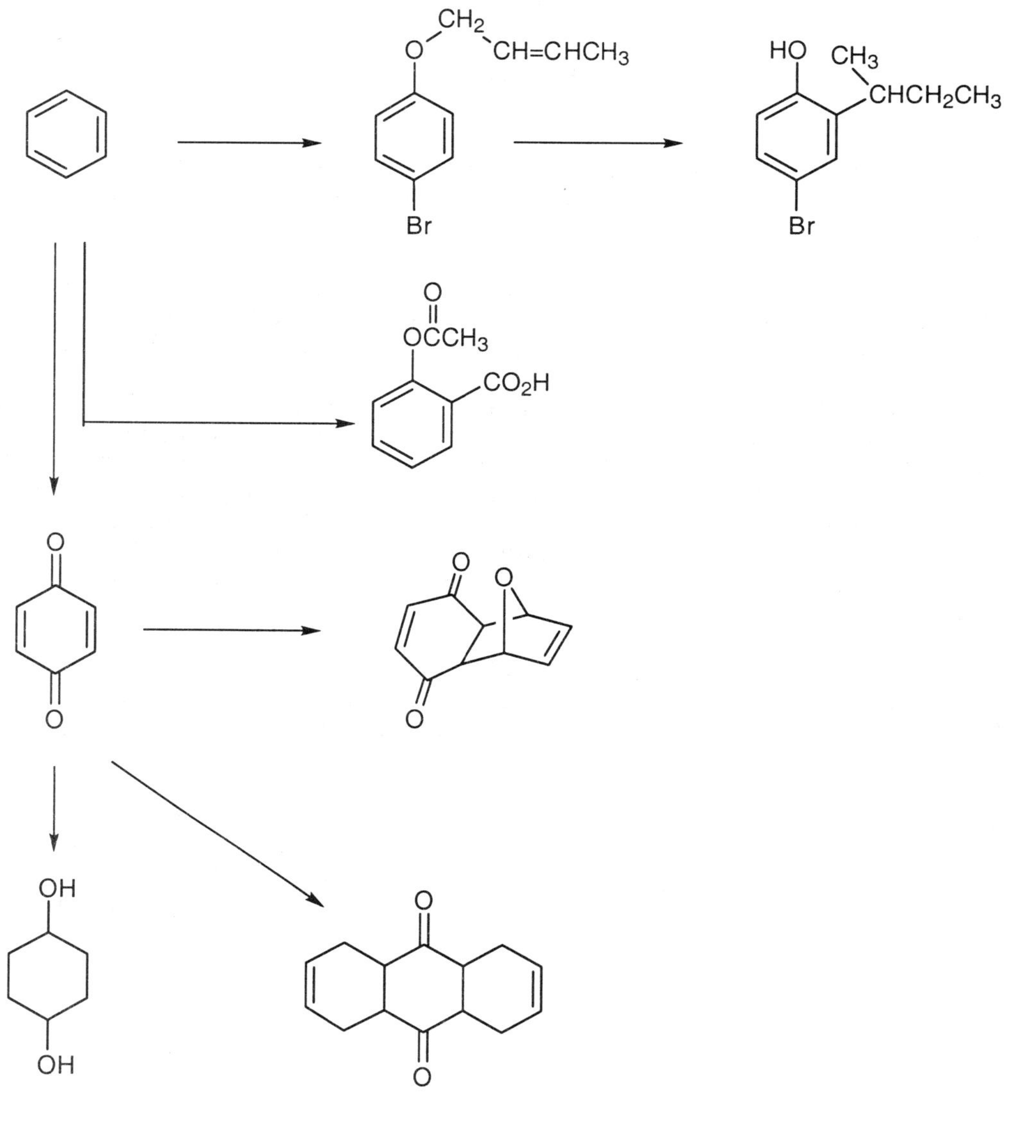

WORKSHOP GUIDE

Week 25

Phenols

The connection between ideas about acidities and experimental observations seems to need continuous repetition and reinforcement. *Problem 1* presents yet another opportunity while emphasizing the difference between alcohols and phenols.

There is a world of difference between knowing reactants, reagents, and products and knowing the mechanism of a transformation. *Problem 2* reveals the difference. If the methods for the preparation of phenols in *2(a)* are memorized, there is no chance of getting *2(b)* correct.

As we have observed in previous Guides, practicing mechanisms is a big help to understanding and learning synthetically useful reactions. *Problem 3* is a good example. This is also a problem that connects back to previous lessons. For example, the air oxidation of cumene is a chain reaction. The leader needs to bring out the connection to previously studied chain reactions. It is also worth reflecting on the special reactivity of cumene; for example, neither toluene nor isobutylbenzene reacts with O_2 under comparable conditions. The acid-catalyzed rearrangement of the hydroperoxide gives a hemiketal and is reminiscent of the reaction of ketones with peracids to give esters.

Problems 3(c) and *3(d)* can be analyzed and learned by thinking about reactivity of phenols and phenolate ions in the context of previous understanding of the reactivity of enols and enolate ions. The names themselves tell us to make the connections! *Problem 3(e)* reveals that the quinone is an α,β-unsaturated ketone.

Finally, *Problem 4* contains some very important C–C bond-making reactions, one of which was encountered before in the section on dienes. Both C–C bond forming processes are examples of electrocyclic reactions. *Problem 4* also incorporates some redox chemistry. It is important to bring out the redox relationships because they define the type of reagent that is required.

WORKSHOP

Week 26

Carbohydrates

1. **Reminder:** Fischer projection is a convention for representing 3-D molecules on 2-D surfaces. As in all conventions, there are specific rules to be obeyed. (*R*)-Glyceraldehyde is the simplest monosaccharide. *Convert the 3-D representation into a Fischer projection formula.*

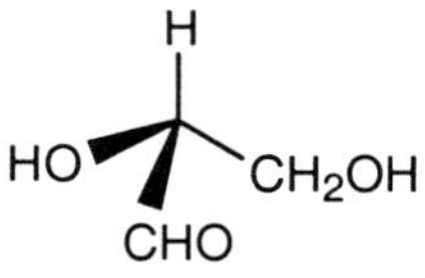

(*R*)-glyceraldehyde

2. Provide structures that satisfy the following descriptions:

   a. The aldopentose that would give the same aldaric acid as D-arabinose upon oxidation with nitric acid.

   b. The aldohexose that gives an optically inactive aldaric acid upon treatment with nitric acid and gives D-ribose on sequential treatment with (i) Br_2, H_2O, and (ii) $Fe_2(SO_4)_3$, H_2O_2.

   c. The enediol that is an intermediate in the isomerization of the carbonyl group of D-arabinose from the 1-position to the 2-position in aqueous base.

   d. A Fischer projection of the L-aldopentose that (i) gives an optically active aldaric acid when oxidized with nitric acid and (ii) is oxidized by sequential treatment with Br_2/H_2O and $Fe_2(SO_4)_3/H_2O_2$ to give a compound that gives an optically inactive alditol upon reduction with sodium borohydride.

3. Provide structures for the major organic products of the following reactions:

   a. 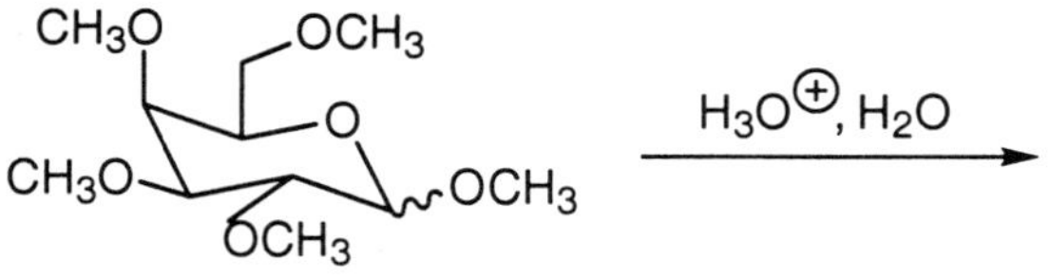

(give Fischer projection)

   b. D-mannose $\xrightarrow[\text{cat } H_3O^{\oplus}]{CH_3CH_2OH}$ (give chair conformation of product(s))

   c. D-glucose $\xrightarrow[H_2O]{NaBH_4}$ (draw Fischer projection)

d. D-glucose

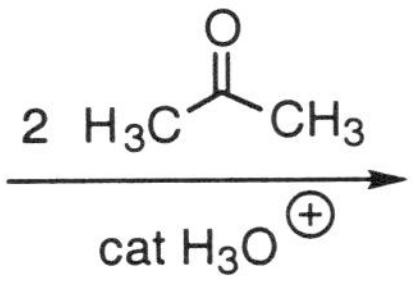

a primary alcohol, $C_{12}H_{20}O_6$ (give chair conformation)

e.

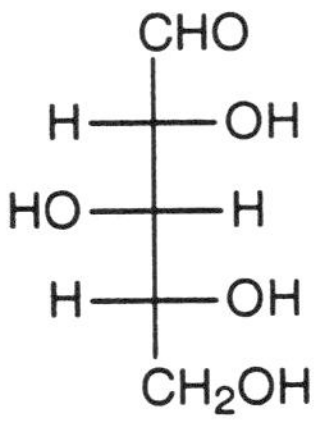

1. HCN, CN^-
2. H_3O^+, H_2O, Δ
3. H_2, Pd($BaSO_4$)
4. H_3O^+, H_2O

4. When D-ribose is treated with methanol and HCl, a mixture of glycosides is formed. Two of these are shown below. Describe with words and chemical equations how periodic acid could be used to distinguish between these compounds.

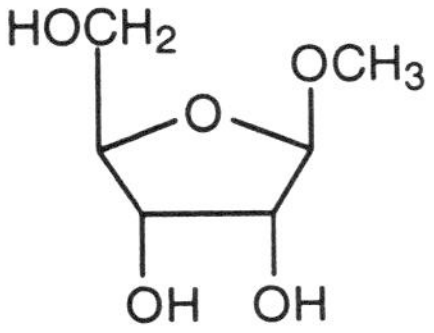

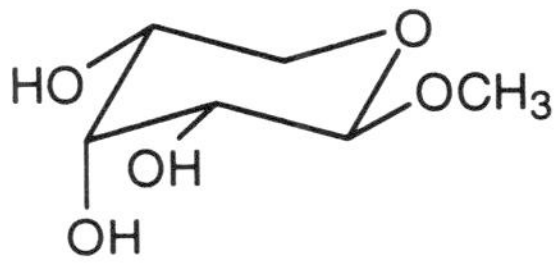

5. Optically active carbohydrate **A** ($C_{14}H_{20}O_7$) does not react with Cu(II) complex and upon treatment with dilute HCl gives optically active carbohydrate **B** ($C_7H_{14}O_7$) plus benzyl alcohol. Reaction of **A** with excess dimethyl sulfate gives **C**, which upon dilute acid hydrolysis gives the optically active 2,3,5,6,7-pentamethyl aldoheptose **D**. When **B** is treated with nitric acid, optically inactive **E** is formed. Oxidative removal of C_1 of **B** gives D-glucose. Give structures **A** through **E** consistent with this information.

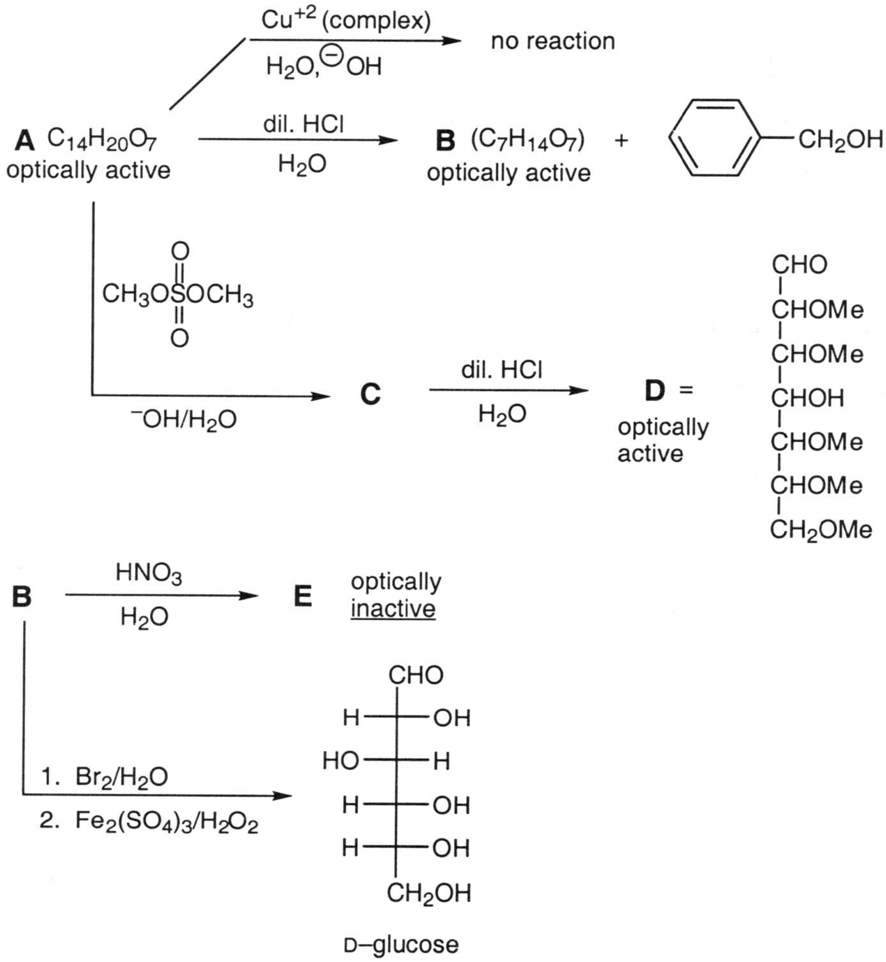

WORKSHOP GUIDE

Week 26

Carbohydrates

So, why did we do all the work on chirality earlier in the course? One of the reasons is so to be able to make sense out of carbohydrate chemistry. The leader will want to make this connection; *Problem 1* is a reminder about CIP nomenclature and asymmetric centers. *Problem 1* also is intended to make sure that everyone is playing by the same Fischer projection rules.

Problem 2 poses short **Observation-Deduction** problems that set the stage for the more complex *Problem 5*. *Problem 4* is an "inverse" **Observation-Deduction** problem that starts with the structures and asks the students to work back to the observations. For the beginning student, much of the residual value from carbohydrate chemistry is the logic of the structure proof. This Workshop emphasizes that logic.

Problem 3 is a quick trip through some important functional group manipulations that are important for synthesis and structure proof. The reactions are not new and connect back to the study of aldehydes. The discussion needs to bring out those connections.

WORKSHOP

Week 27

Review

1. Make lists showing as many examples as you know of reactions that involve each of the following.

   a. Formation of new C–C bonds

   b. Carbanion intermediates

   c. Carbocation intermediates

   d. Reduction of organic molecules

   e. Oxidation of organic molecules

   f. Electrophilic addition to π bonds

   g. Nucleophilic addition to π bonds

   (*Hint*: It may be fun to do this problem in a round-robin format.)

2. Consider the following problems about rates and equilibria:

   a. Benzyne and anthracene react to produce the following products:

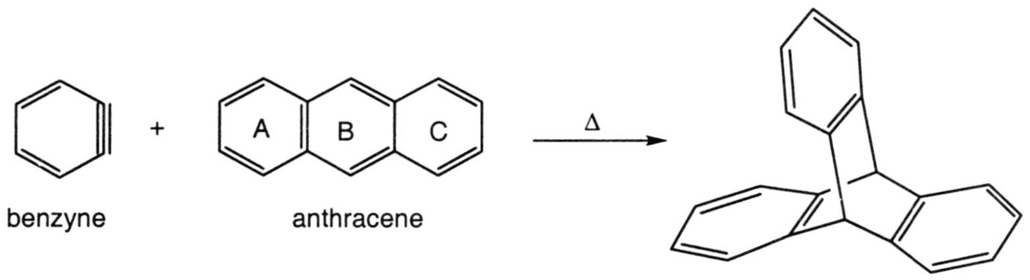

Suppose that benzyne also added to ring C. What would be the structure of the ring C adduct? Analyze the structures of the ring B and C adducts and suggest an explanation for the observation that ring B adduct is the only product.

b. Rationalize the observed equilibria.

i. 

versus

ii. 

versus

c. Treatment of the following compound with iodomethane led to the formation of two isomeric products, **A** and **B**. When each of these was treated first with Ag_2O/H_2O and then heated, the same mixture of compounds **C** and **D** were formed. Compound **C** could be resolved into enantiomers, but **D** could not.

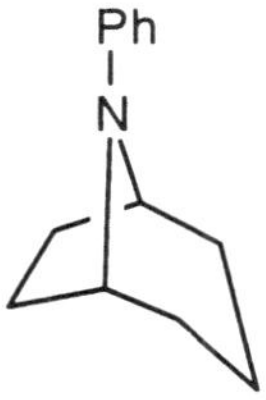

i. Provide structures for **A** and **B** and name the structural relationship between them.

ii. Provide structures for **C** and **D**. Predict the approximate product composition: **C/D** >> 1; << 1; ~1.

d. Arrange the following molecules in order of *decreasing* rate of reactivity toward excess aqueous 0.1 N NaOH solution

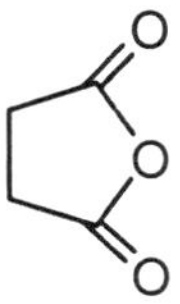

E

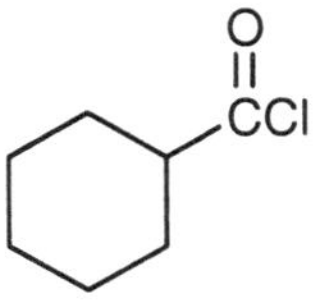

F

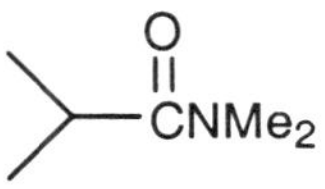

G

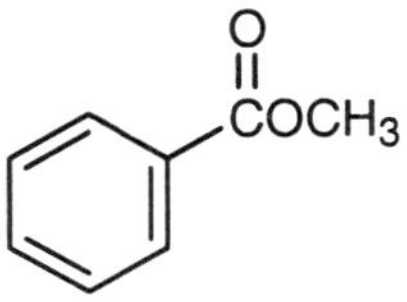

H

e. Arrange the following molecules in order of decreasing rate of exchange with $H_2{}^{18}O$ in the presence of 0.01 M H_3O^+.

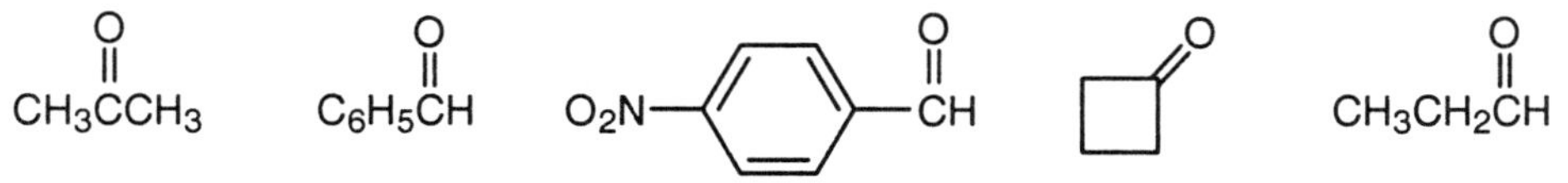

3. Consider each of the following reactions. First, decide whether the reaction is net oxidation, net reduction, or neither. Second, specify an appropriate reagent to bring about the transformation. Specify the half-cell for the redox reagents.

a. 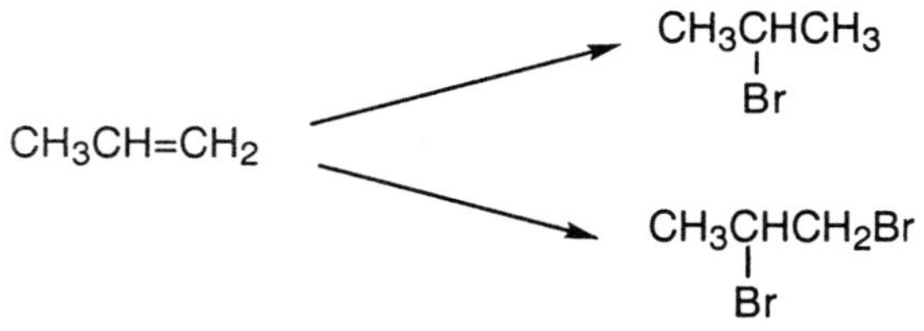

b. 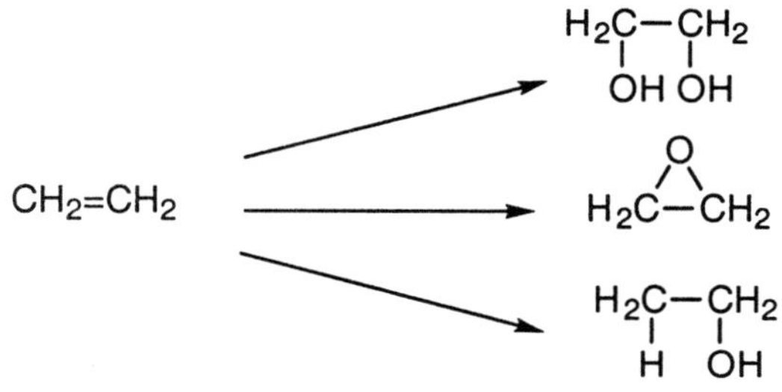

c. 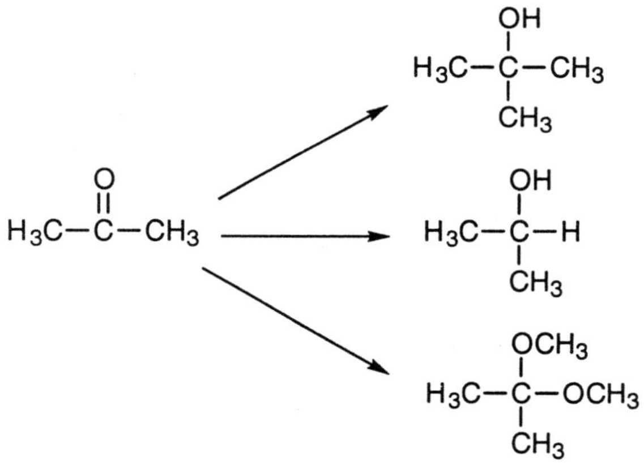

d. $H_2C{=}CHCH_2Br \longrightarrow H_2C{=}CHCH_2MgBr$

e. 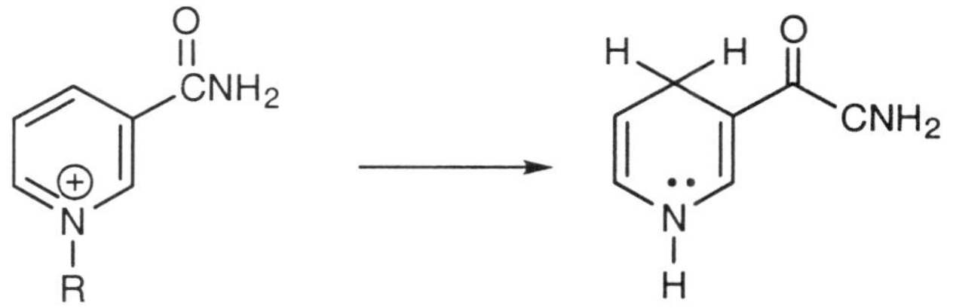

f. 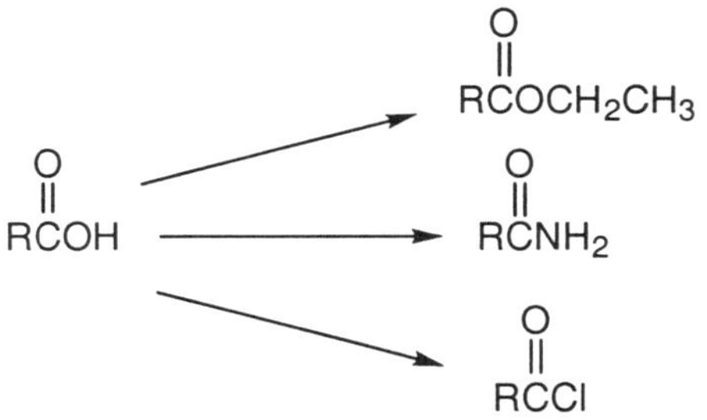

g. 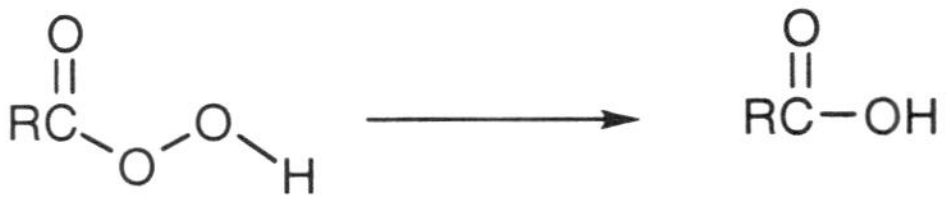

h. 

i. 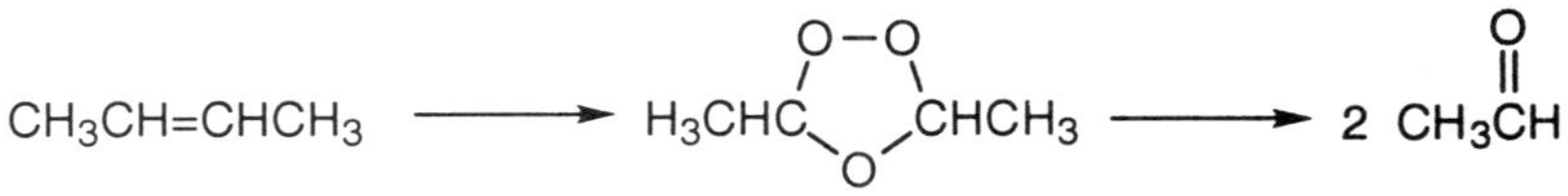

4. The following reactions do not occur, or occur slowly, under ordinary conditions in the absence of a catalyst. Suggest the necessary catalyst that makes these reactions useful. For each reaction, indicate the new (lower activation energy) mechanistic pathway that is opened up by the catalyst.

a. $CH_2{=}CH_2 + H_2 \longrightarrow H_3C{-}CH_3$

b. $CH_2{=}CH_2 + H_2O \longrightarrow CH_3CH_2OH$

c. 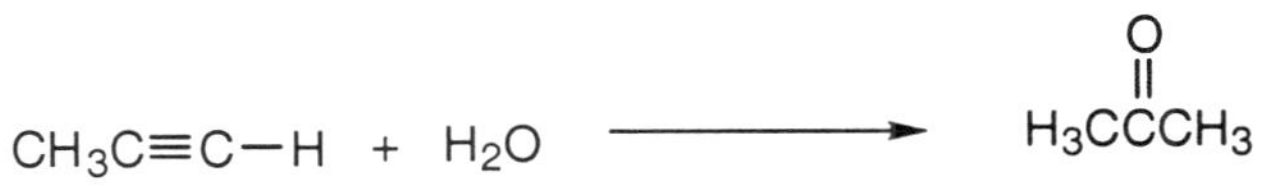

d. $(CH_3)_3CH + Br_2 \longrightarrow (CH_3)_3CBr + HBr$

e. $C_6H_6 + Br_2 \longrightarrow C_6H_5Br + HBr$

f. 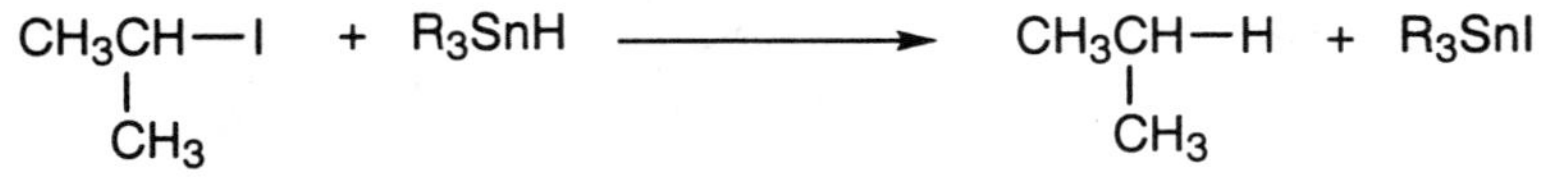

g. 

h. $CH_3CH_2{}^{35}Cl + Na^{37}Cl^{\ominus} \longrightarrow CH_3CH_2{}^{37}Cl + {}^{35}Cl^{\ominus}$

i. 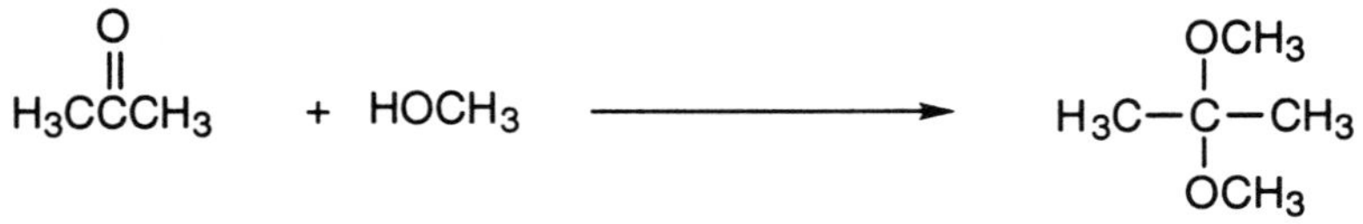

j. 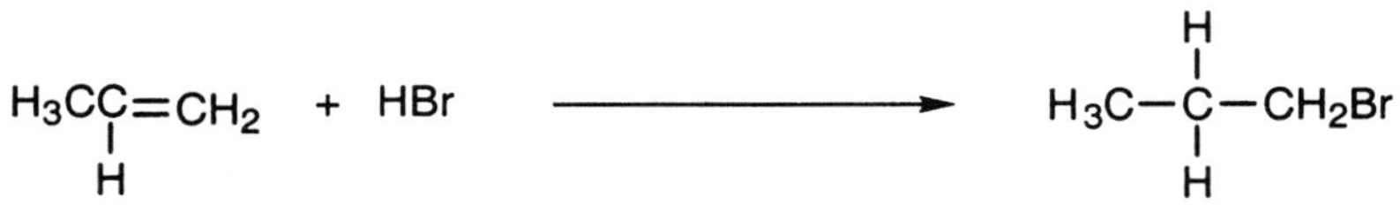

k. 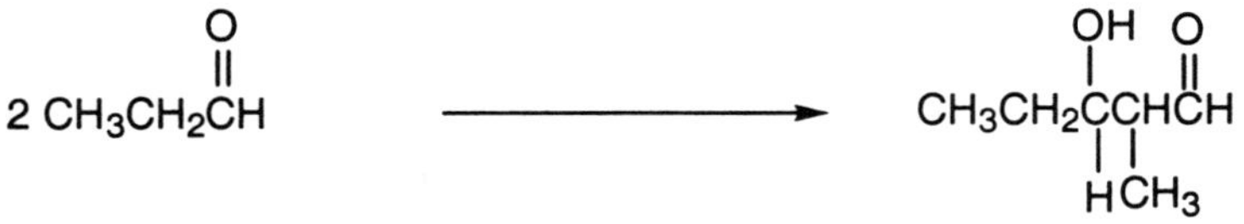

l. 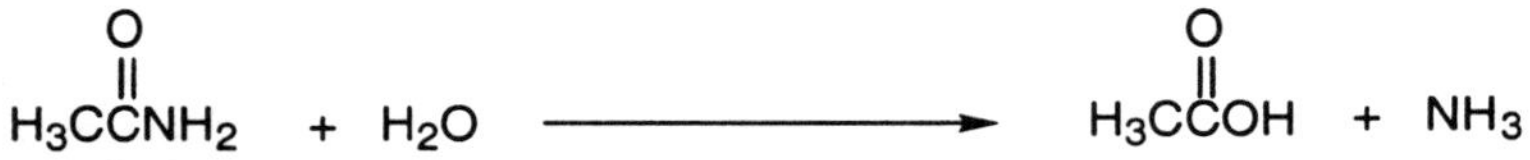

m. $(CH_3)_3COC_6H_5 + H_2O \longrightarrow (CH_3)_2C{=}CH_2 + HOC_6H_5$

n. $C_6H_6 + C_6H_5COCl \longrightarrow C_6H_5COC_6H_5 + HCl$

o. 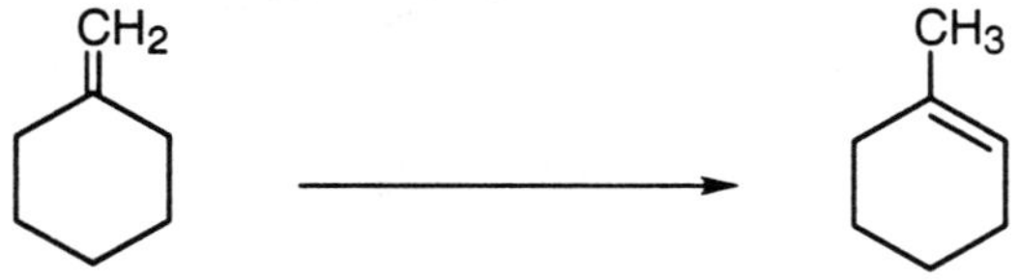

p. 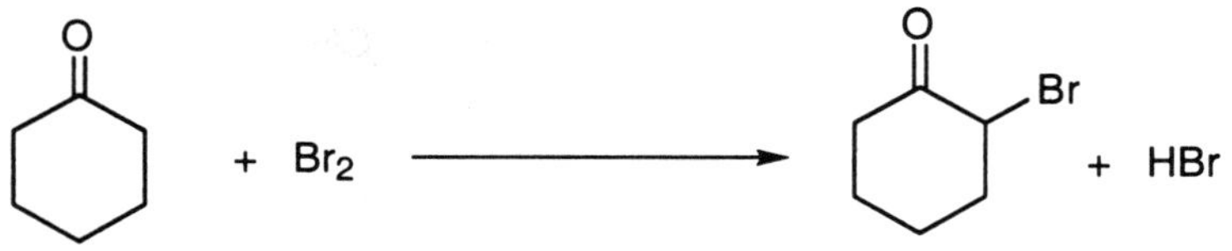

q. 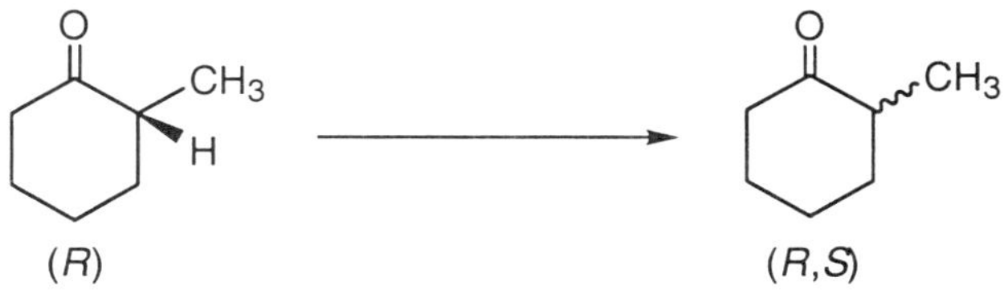

r. $2\ C_6H_5CHO \longrightarrow C_6H_5C(=O)CH(OH)C_6H_5$

s. $(CH_3)_3CH + D_2SO_4 \longrightarrow (CD_3)_3C\text{–}H + H_2SO_4$

t. 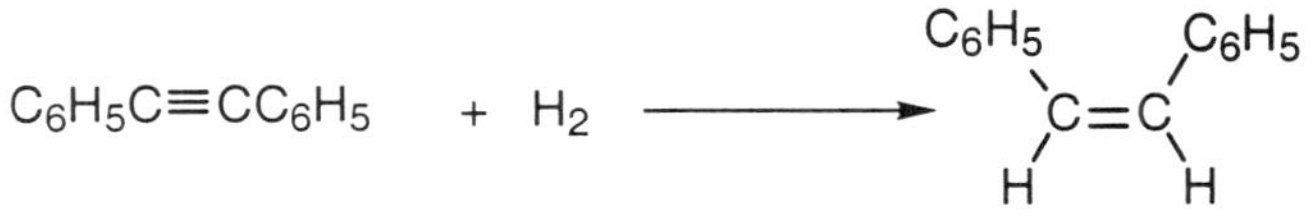

u. $C_6H_5N_2^{\oplus} + CN^{\ominus} \longrightarrow C_6H_5CN$

v. $PhCHO + CH_3NO_2 \longrightarrow PhCH{=}CHNO_2 + H_2O$

w. $CH_3CH{=}C{=}CH_2 \longrightarrow CH_3CH_2C{\equiv}CH$

x. 

y. $t\text{–BOCNHCH(CH}_3\text{)CO}_2\text{H} + H_2NCH_2CO_2H \longrightarrow t\text{–BOCNHCH(CH}_3\text{)C(=O)NHCH}_2\text{CO}_2\text{H}$

z. $C_6H_5CHO + (CH_3)_2CHOH \longrightarrow C_6H_5CH_2OH + (CH_3)_2C{=}O$

5. a. Propose a structure for disaccharide **A** and indicate what feature of the structure is not determined by the evidence provided. Present your structure using careful conformational drawings; mark the stereochemistry of substituents clearly and unambiguously as axial (ax) or equatorial (eq) on the conformational drawing.

An unknown disaccharide **A**, $C_{11}H_{20}O_{10}$, does *not* react with Tollen's reagent to give a silver mirror. Hydrolysis of **A** with aqueous acid gives D-glucose and D-ribose. A β-glucosidase has no effect on **A**, but an α-glucosidase hydrolyzes **A** to D-glucose and D-ribose. When **A** is methylated [$(CH_3O)_2SO_2$, NaOH], there is obtained a heptamethyl ether **B** that can be hydrolyzed with aqueous acid to 2,3,4,6-tetra-O-methyl-D-glucose and 2,3,4-tri-O-methyl-D-ribose.

b. Provide structures for **A**, **B**, and **C** that are consistent with the following observations.

Compound **A**, $C_7H_{15}NO$, was neutral and exhibited strong absorption in the IR spectrum at 1650 cm^{-1} but none in the 3200-3400 cm^{-1} region. Extended heating of **A** with H_3O^+/H_2O gave compound **B** and, after addition of NaOH, a nitrogen-containing compound **C**. **B** exhibited strong absorption in the IR spectrum at 1712 cm^{-1} and a very broad absorption over the range 2500-3500 cm^{-1}. The ^1H NMR spectrum of **B** consisted of the following absorptions: δ 1.20 (d, rel. area 6, J = 7.0 Hz), 2.41 (septet, rel. area 1, J = 7.0 Hz), and 12.40 (s, rel. area 1). Compound **C** exhibited a single absorption at 3335 cm^{-1} in the IR spectrum and reacted with nitrous acid at 0 °C to give a yellow oil. Compound **C** was synthesized by treating ethyl amine first with formic acetic anhydride and then with $LiAlH_4$ ($H_3O^{\oplus}$ workup).

c. Give structures for **B** through **E** consistent with the following information:

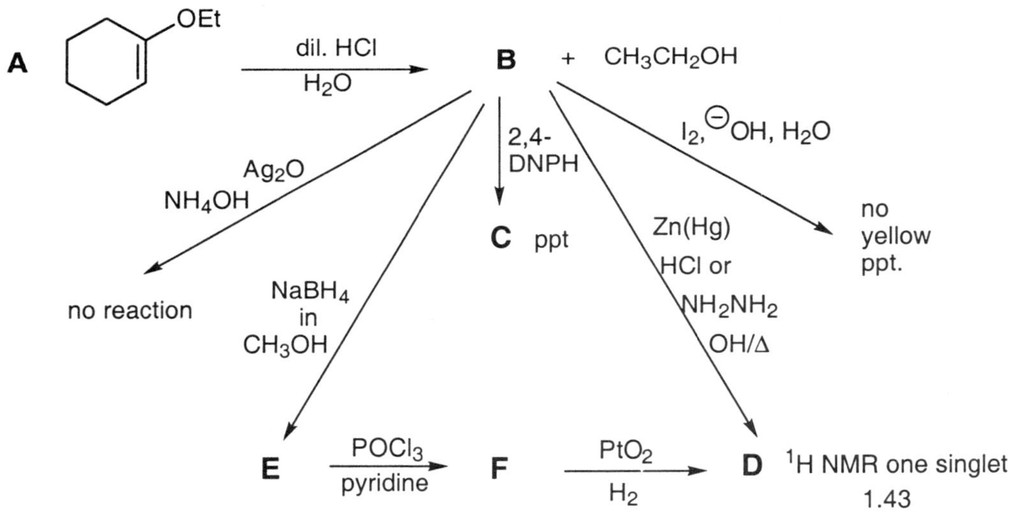

6. Provide structures for the major product(s) of the following reactions. Explain with words and structures your choice of the major product.

a. 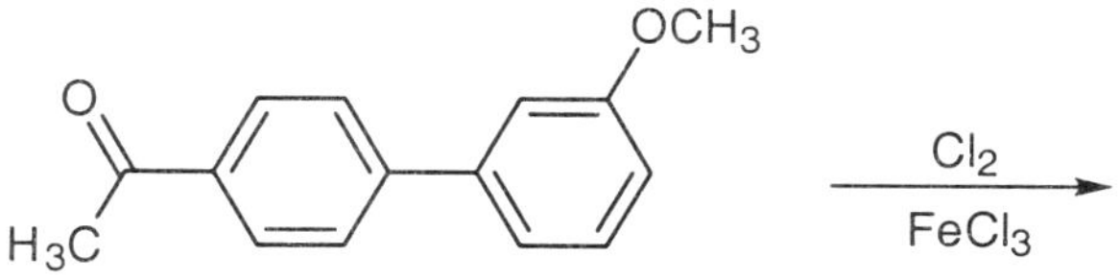

b. 

c. 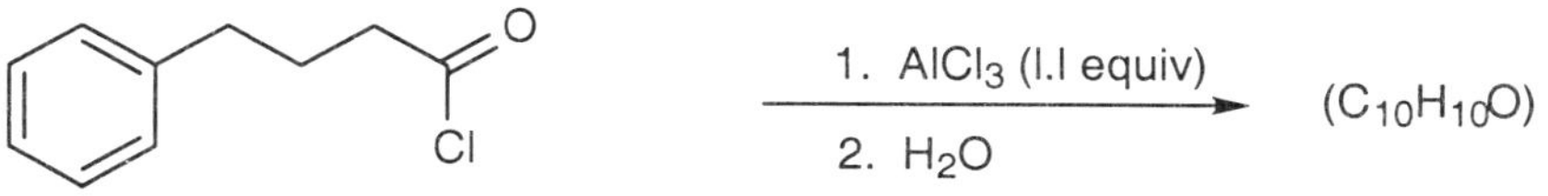

d. 

e. 

f. 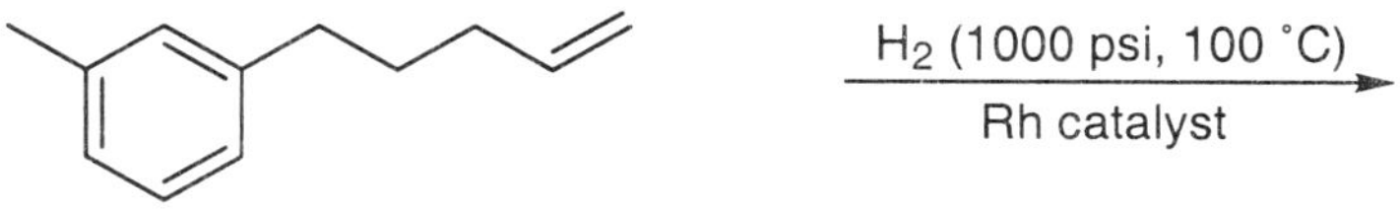

g. 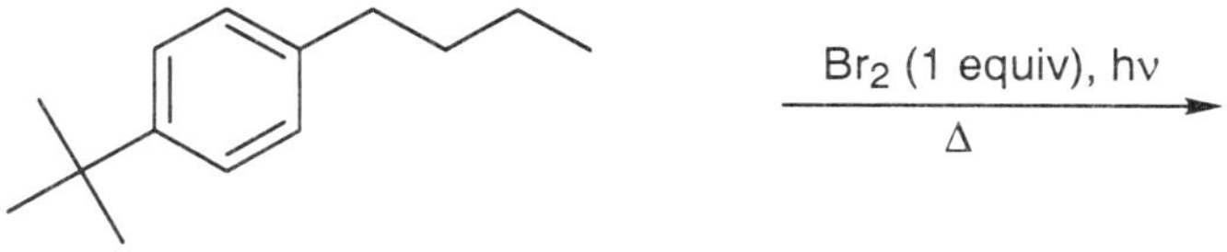

h. 

i. 

7. "Disconnect" the following compounds into two fragments by breaking appropriate C–C bonds. Choose the C–C bond carefully so that the reaction of the two fragments would reassemble the C–C bond.

a. 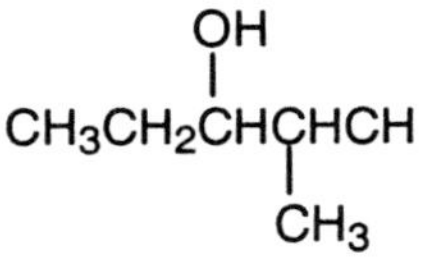

b. 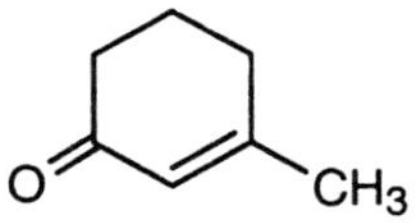

c. 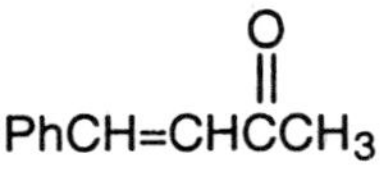

d. 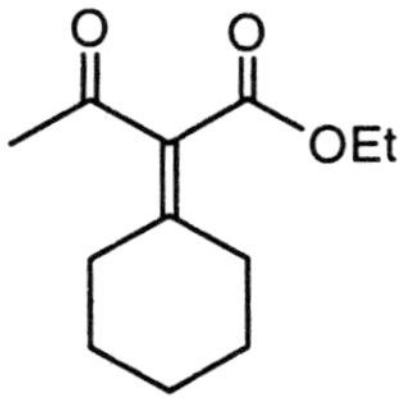

e. 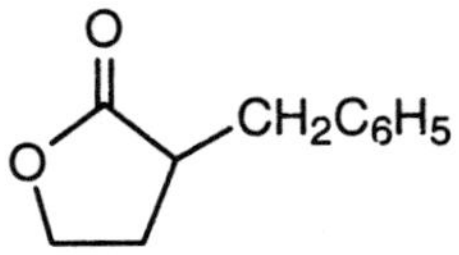

f. 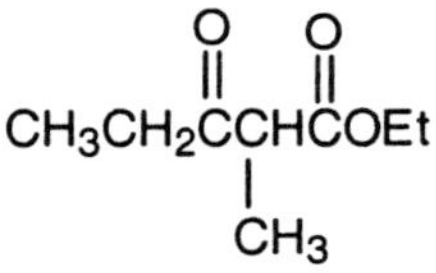

g. 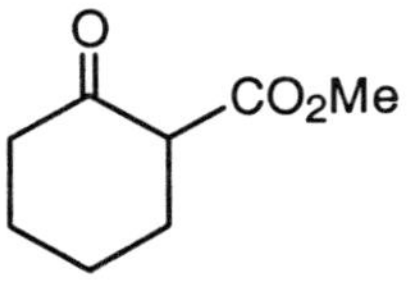

h. 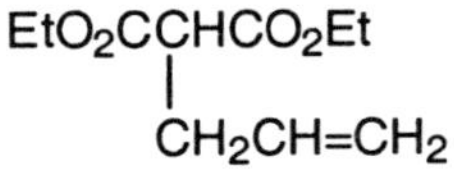

i. 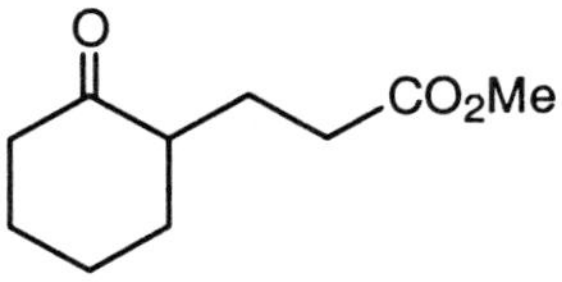

j. 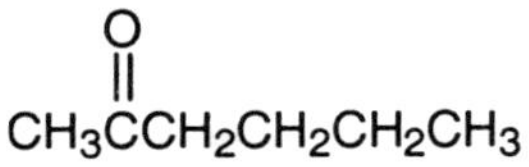

k. 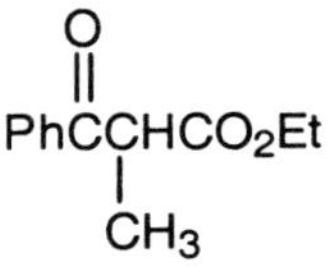

l. 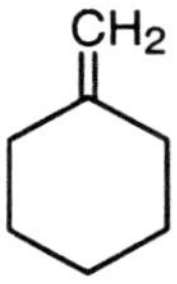

8. a. Show how to carry out the indicated synthetic conversions in an efficient manner. If more than one step is required, show the products for each step. Make as many synthetic connections as possible with compounds on the periphery of the "starburst."

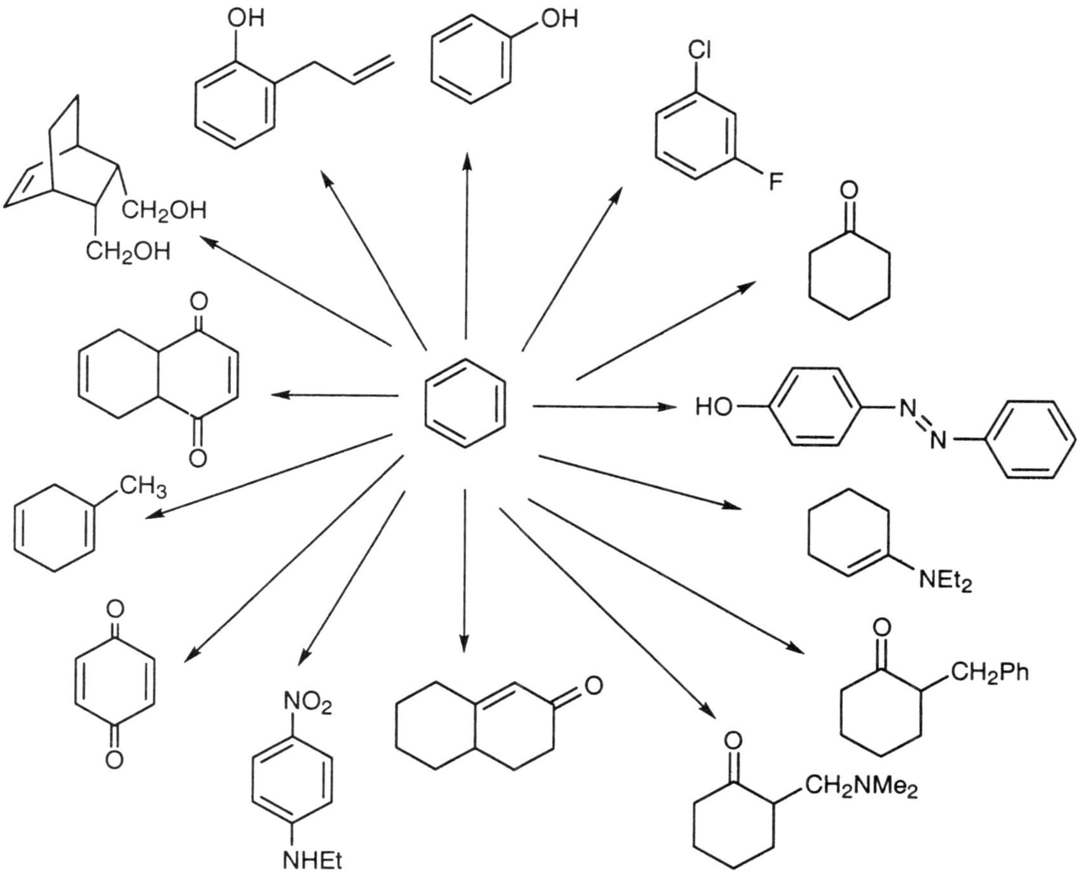

b.

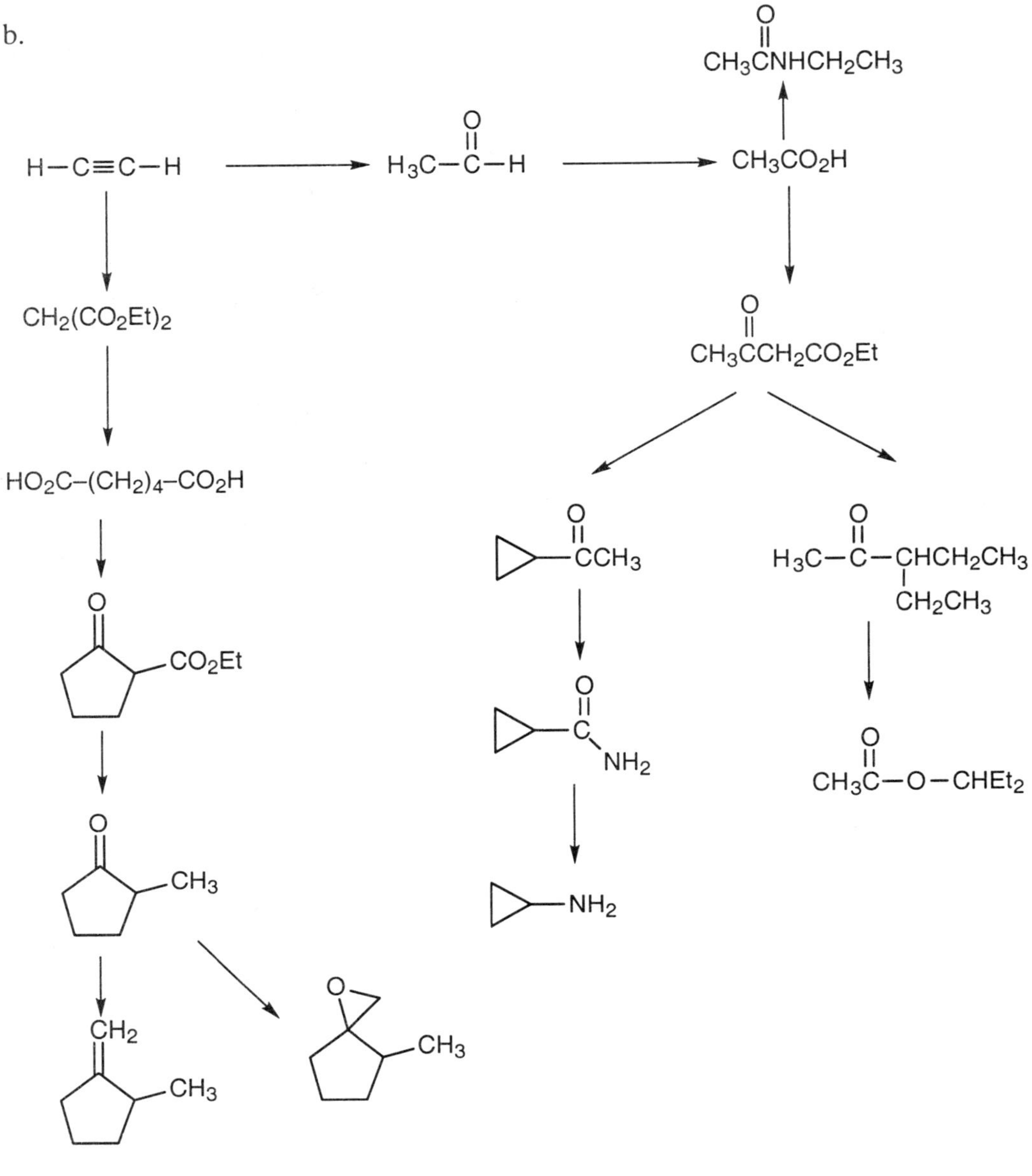
H–C≡C–H
H3C–C(=O)–H
CH3CO2H
CH3C(=O)NHCH2CH3
CH2(CO2Et)2
CH3C(=O)CH2CO2Et
HO2C–(CH2)4–CO2H
CO2Et
CCH3
H3C–C(=O)–CHCH2CH3
CH2CH3
NH2
CH3C(=O)–O–CHEt2
CH3
O
CH2

WORKSHOP GUIDE

Week 27

Review

This Workshop is designed as an open-ended review. Clearly, there is too much work for a single Workshop. The leader might choose a few examples from each type of problem. The idea is to get the students started in the Workshop and stimulated to continue to work on the review problems with friends and neighbors.

Reviews work best if they help students organize and make connections between observations and ideas. Any problem that gets students working with the material in new ways will be helpful. Thus, the review problems constitute a guided tour through the **big ideas**: mechanism, structure-rate, structure-equilibrium, redox, catalysis, structure proof, functional group reactivity, retrosynthetic analysis, and synthesis.